ELMHURST PUBLIC LIBRARY
3 1135 00552 3615

W9-DES-281

YOUNG SCIENTIST CONCEPTS & PROJECTS

WEATHER

ROBIN KERROD

Gareth Stevens Publishing
MILWAUKEE

ɔriginal publishers would like to thank the following pupils from St. John Baptist Church of England School and Walnut Tree Walk Primary School: Maria Bloodworth, Tony Borg, Steven Briggs, Jackie Ishiekwene, Daniel Johnson, Jon Leaning, Erin Macarthy, Hanife Manur, Tanya Martin, Lola Olayinka, and Ini Usoro. Thanks to West Meters Ltd. for the loan of props.

For a free color catalog describing Gareth Stevens' list of high-quality books and multimedia programs, call 1-800-542-2595 (USA) or 1-800-461-9120 (Canada). Gareth Stevens Publishing's Fax: (414) 225-0377. See our catalog, too, on the World Wide Web: http://gsinc.com

Library of Congress Cataloging-in-Publication Data

Kerrod, Robin.
Weather / by Robin Kerrod.
p. cm. — (Young scientist concepts and projects)
Includes bibliographical references and index.
Summary: Discusses such aspects of weather as wind, rainbows, and climate and presents a variety of experiments and other activities involving the weather.
ISBN 0-8368-2088-6 (lib. bdg.)
1. Weather—Juvenile literature. 2. Meteorology—Juvenile literature. 3. Climatology—Juvenile literature. [1. Weather. 2. Meteorology. 3. Weather—Experiments. 4. Meteorology—Experiments. 5. Experiments.] I. Title. II. Series.
QC981.3.K47 1998
551.5—dc21 97-41675

This North American edition first published in 1998 by
Gareth Stevens Publishing
1555 North RiverCenter Drive, Suite 201
Milwaukee, WI 53212 USA

Senior Editors: Sue Grabham and Caroline Beattie
Editor: Sam Batra
Photographer: John Freeman
Stylist: Thomasina Smith
Designer: Caroline Reeves
Picture Researcher: Liz Eddison
Illustrator: Michael Lamb
Gareth Stevens series editor: Dorothy L. Gibbs
Editorial assistant: Diane Laska

Printed in the United States of America

1 2 3 4 5 6 7 8 9 02 01 00 99 98

YOUNG SCIENTIST CONCEPTS & PROJECTS

WEATHER

CONTENTS

WATCHING WEATHER

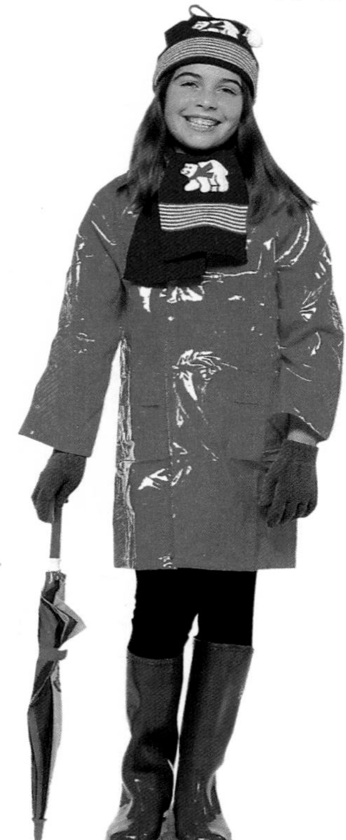

THE weather affects us all. If it is warm and sunny, we like to spend time outside and wear light clothes. If it is cold and wet, we prefer to stay indoors. If we do go out, it is best to dress in layers and carry an umbrella. What exactly does the word *weather* mean? It means the conditions in the air around us – how hot it is, how strong the wind is, whether it is sunny or cloudy, whether it is dry or wet. Because the weather is so important, thousands of scientists around the world study it. The scientific study of the weather is called meteorology, and the scientists who study it are called meteorologists. One of their main jobs is to forecast the weather, that is, tell us what the weather is going to be like in the future. However, their forecasts are not always right!

It is winter, and the weather forecast says it is going to be cold and wet. You need to put on warm, waterproof clothes. Remember an umbrella!

Snowball fights are a lot of fun as long as the snowballs don't hit you. Snow is fun to play in, but it can make traveling very dangerous.

In the summer, the sun is so strong that you need to wear a hat and sunglasses. The coolest place to be is in the water, but, even there, you need protection from the sun. Anyone for a swim?

Learning about the sun

Floating high above the earth, in space, a satellite called *Solar Max* keeps a close eye on the sun. The results it sends back help scientists understand how the sun affects the world's weather and climate.

This weather vane tells us that a west wind is blowing. A west wind is one that blows from the west, not toward the west.

Frozen over

Changes in the sun's heat output could have caused the Little Ice Age that occurred in the 1600s. During this time, the Thames River froze so hard during winters that events called Frost Fairs were held on the river in London, England. This picture *(above)* shows the Frost Fair of 1683.

THE SUN'S ENERGY

Core

Energy flow

THE earth gets almost all of its heat from the sun. The sun is a star. Like other stars, the sun releases tremendous amounts of heat, light, and other forms of energy into space. Only a small amount of the sun's energy reaches the earth, but it is enough to power our weather systems. Its energy can make rocks so hot that you can fry eggs on them. It can create tornadoes that toss cars high into the air. Varying amounts of the sun's heat fall on different parts of the world, which is the main reason why places around the world have different weather. The amount of heat a place receives from the sun also changes with the seasons.

The sun's energy is produced in its center, or core. There, the temperature reaches about 27 million degrees Fahrenheit (15 million degrees. Celsius). At this temperature, atoms of gas fuse, or combine, and give out enormous amounts of energy. This energy escapes into space, mainly as light and heat.

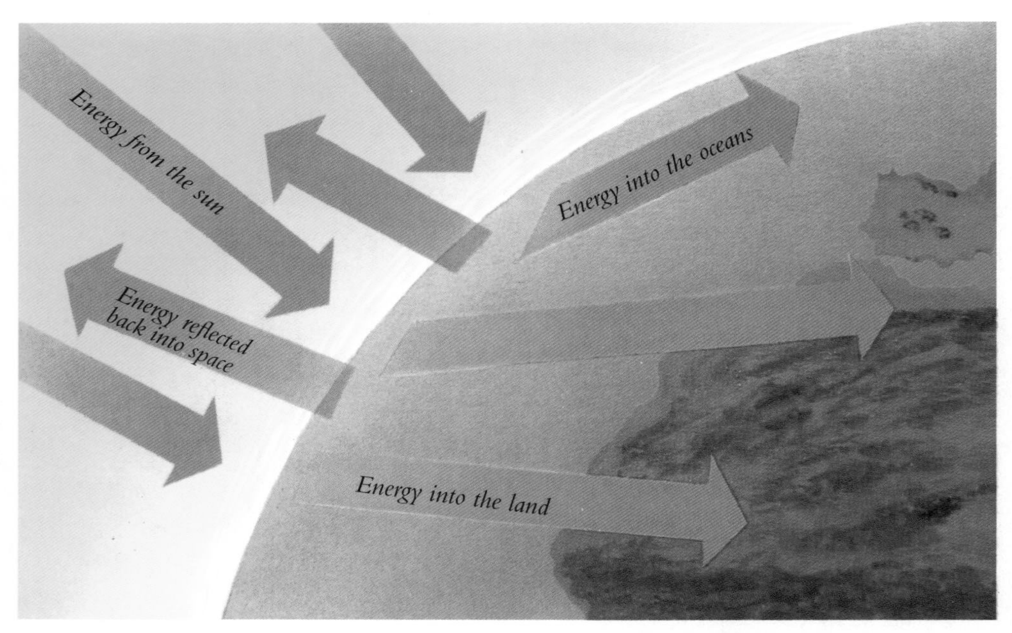

Heating the earth
The sun pours energy onto the earth. Some of it bounces off the atmosphere back into space. Some heats up the air, but most of it heats up the ground and the oceans.

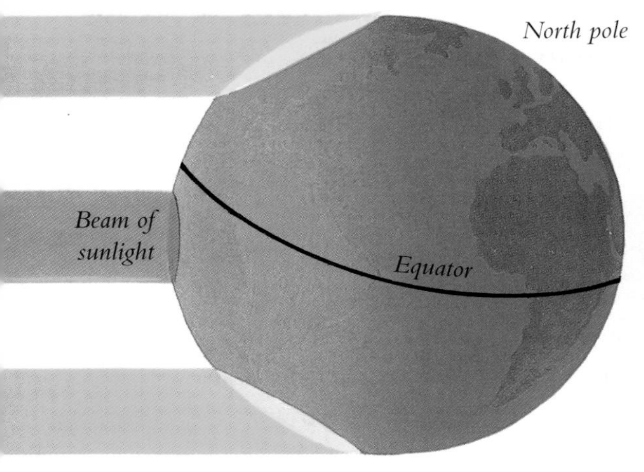

Cooler or warmer?
The sun's energy does not fall evenly over the earth because the earth is round. Imagine three beams of sunlight (*above*), of the same size and with the same amount of energy, falling on the earth. One falls on the equator and the others on the north and south poles. Because the beam falling on the equator covers a much smaller area, its energy is more concentrated and the temperature is higher.

Sunlight into electricity
Solar power plants capture the sun's energy and turn it into electricity. This huge plant in California uses hundreds of mirrors to reflect sunlight onto a boiler at the top of the tower.

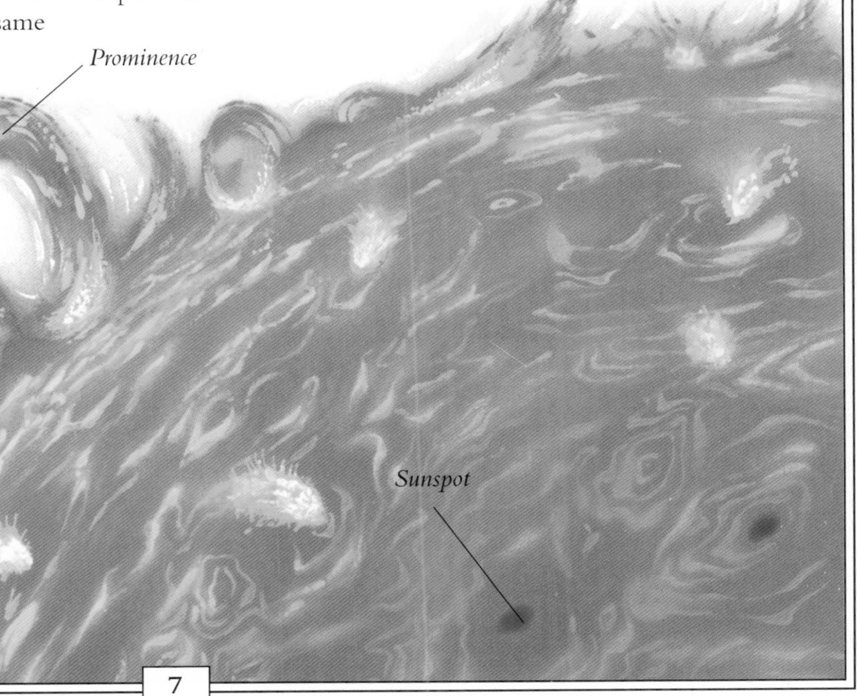

The surface of the sun is a stormy sea of bubbling, boiling gases. In some places, great fountains of gas, called prominences, shoot high above the surface. The temperature of most of the surface is about 10,000°F (5,500°C). Dark patches, called sunspots, are cooler.

MEASURING TEMPERATURE

This wall thermometer is filled with blue-colored alcohol. It has two temperature scales: degrees Celsius (°C) and degrees Fahrenheit (°F).

WHAT we notice most about the weather is the temperature – how hot or how cold it is. We measure the temperature with a *thermometer*, a word that literally means heat measurer. Ordinary thermometers have a column of liquid in a glass tube. When the temperature goes up, the liquid expands, and the column of liquid gets longer. The amount the column lengthens is a measure of how much the temperature has increased. So, thermometers can be used to show a broad range of temperatures. The liquids used most in thermometers are mercury and alcohol. They are used because they do not boil or freeze at temperatures found in the home. But we can make a simple thermometer using water.

Thermo-strips

You can take your temperature by pressing a thermo-strip against your forehead. The strip that shows your temperature changes color. This strip *(above)* indicates 103°F (39.4°C), but usually it registers 98-99°F (36.7-37.2°C), which is the normal body temperature for human beings.

FACT BOX

- A pleasant room temperature for human beings is 75°F (24°C).
- In 1922, at a place called Al´Aziziyah, in Libya, the temperature in the shade rose to 140°F (60°C).
- Water freezes when the temperature falls to 32°F (0°C).
- During winter in parts of North America and northern Europe, temperatures can fall to more than minus 40°F (-40°C).
- At a temperature of about 374°F below freezing (-225°C), air turns into liquid.

MATERIALS

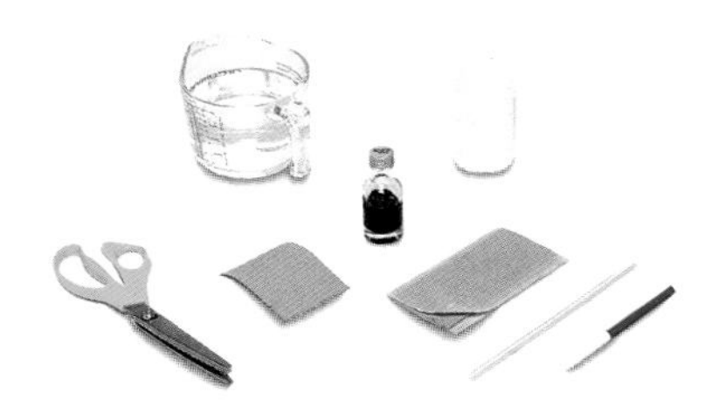

You will need: pitcher of cold water, bottle, food coloring, clear straw, adhesive, small piece of cardboard, scissors, felt-tip marker.

Make a thermometer

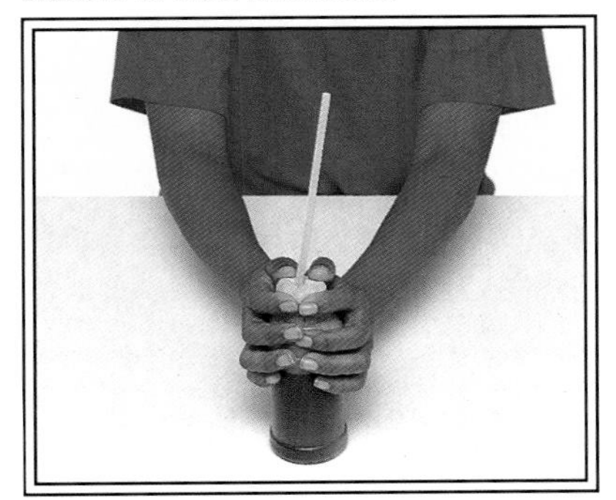

1 Pour cold water into the bottle until it is about two-thirds full. Add some food coloring. Dip the straw into the water and seal the neck tightly with adhesive.

2 Blow into the straw to force extra air into the bottle. After a few seconds, the extra air pressure inside the bottle will force the water to move up inside the straw.

3 Cut slots in the top and bottom of the cardboard and slide it over the straw, as in the picture *(above)*. After the bottle stands for a while, make a mark on the cardboard next to the water level to show the temperature of the room. Then, take your thermometer outside in the sun – that is, if the sun is shining!

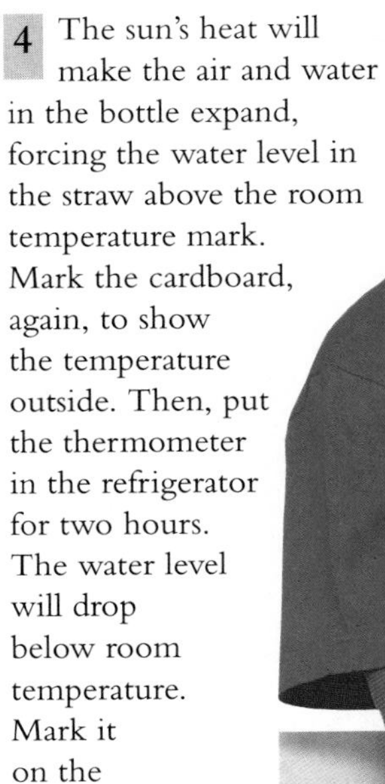

4 The sun's heat will make the air and water in the bottle expand, forcing the water level in the straw above the room temperature mark. Mark the cardboard, again, to show the temperature outside. Then, put the thermometer in the refrigerator for two hours. The water level will drop below room temperature. Mark it on the cardboard.

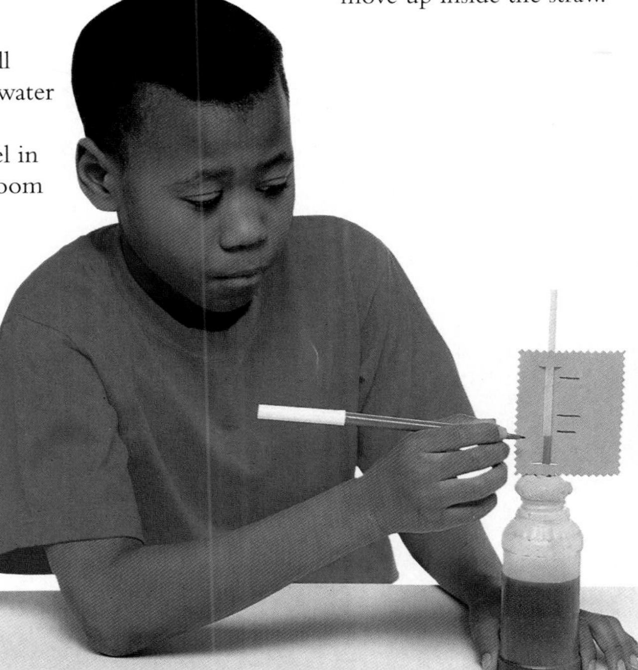

THE ATMOSPHERE

THE air around us forms a layer that covers the earth. We call this layer the atmosphere. The atmosphere is most dense, or thickest, near the ground. It gets less dense, or thinner, the higher up you go. At a height of about 186 miles (299 kilometers), there is scarcely any air left at all. This is the beginning of space. Most of our weather takes place in the lowest and thickest part of the atmosphere, in a layer that we call the troposphere. This layer is 6 to 10 miles (10 to 16 km) thick. It is in this layer that clouds form, rain and snow fall, and thunder and lightning take place. In the next part of the atmosphere, called the stratosphere, there is a layer of a gas called ozone. Ozone is very important to us because it blocks dangerous rays coming from the sun.

Working in space above the atmosphere, astronauts must wear spacesuits to survive. These spacesuits supply them with oxygen to breathe and protection from the sun's heat.

Space
The upper edge of the atmosphere is about 100 miles (161 km) above the earth. Then space begins. Above the atmosphere is the moon, nearly 238,700 miles (384,068 km) away.

Heavenly bodies
The thin streak in this picture *(above)* was made by a meteor, or shooting star, which is a small piece of rock that burns up when it plunges into the atmosphere. The broader trail was made by a comet, which is a huge lump of icy rock that gradually breaks up in space. Occasionally, we can see comets in the sky.

Lights in the sky
Strange glowing patches appear in the skies above Alaska. They also can be seen in areas around the North and South poles. In the north, they are called the Northern Lights, or the Aurora Borealis. In the south, they are called the Southern Lights, or the Aurora Australis.

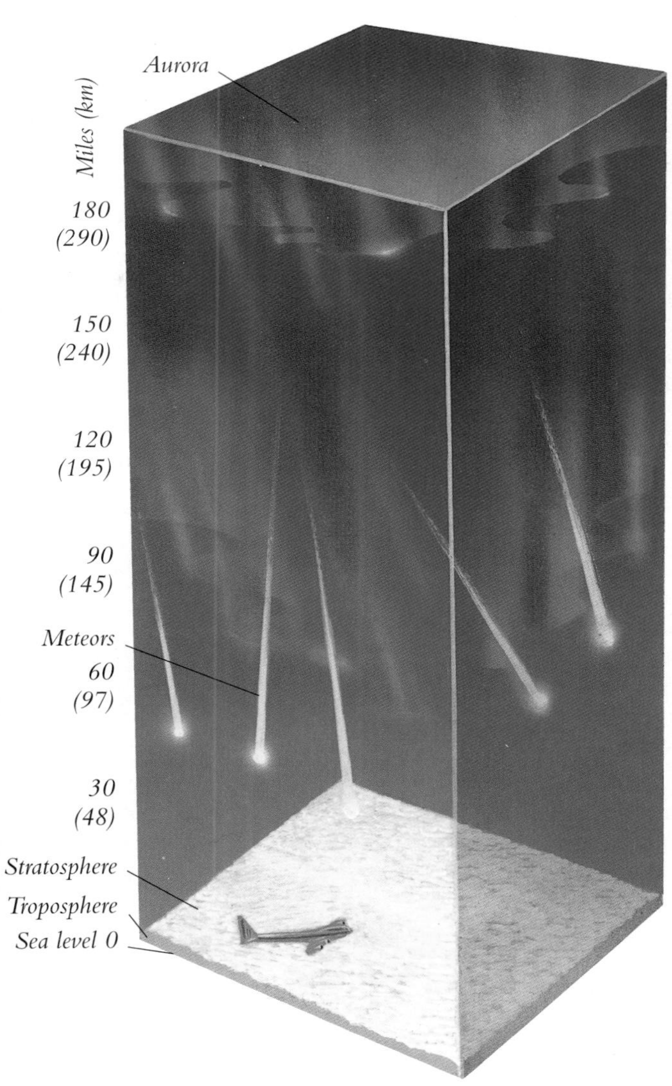

A section of the earth's atmosphere shows its various layers.

AIR AND SKY

THE air that makes up the atmosphere is a gas that we cannot see, feel, taste, or smell. Actually, it is not one gas, but a mixture of many different gases. The two main gases are nitrogen and oxygen. The chart *(below, left)* shows that there is nearly four times as much nitrogen as oxygen in the air. To living things, however, oxygen is the most important gas. Almost all living things must breathe in oxygen to stay alive. The air also contains small amounts of other gases. One of them is carbon dioxide. Plants take in carbon dioxide to make their food, and animals give off carbon dioxide when they exhale. Much of the carbon dioxide in the air comes from cars and factories that burn fuels like oil and coal.

Paragliding is a seaside sport. Wearing a parachute, you are towed high into the air behind a boat. Then you float to the ground.

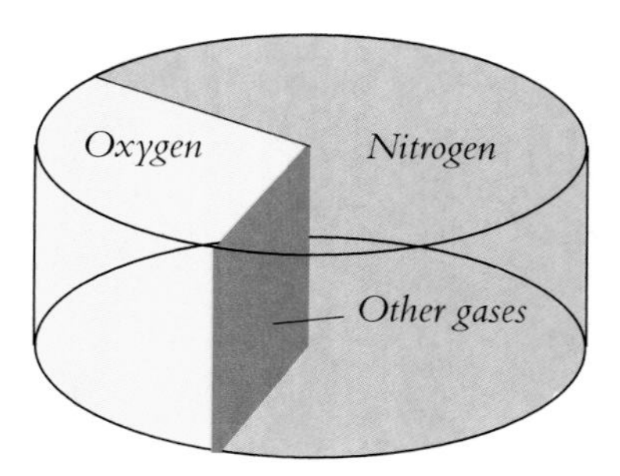

This pie chart shows the amounts of the main gases that make up the air.

Blue sky
The sky looks blue because the air particles scatter more blue light into our eyes than any other color.

Carried in the air
The air contains minute traces of many different substances, such as the scent and pollen of flowers.

Plants and animals
Plants and animals affect the air in different ways. Plants take in carbon dioxide from the air and give off oxygen. Animals do the opposite. They take in oxygen from the air and give off carbon dioxide.

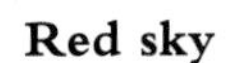

Red sky
In the evening, the sky often turns orange or red, because dust in the lower air blocks the blue rays in the sunlight when the sun is low. Only orange and red rays are able to pass through the dust.

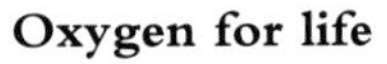

Oxygen for life
Like all animals, we get our energy by burning food inside our bodies. We take in the oxygen to burn our food when we breathe air into our lungs.

IN THE AIR

As we have seen, air is made up of a mixture of different gases, mainly nitrogen and oxygen. We can find out approximately how much oxygen is in the air with this simple experiment. The oxygen inside a jar can be removed by burning a candle in the jar. The candle, or anything else, needs oxygen to burn. During this experiment, water rises in the jar to take the place of the oxygen. By noting how much the water rises, you easily can figure out how much oxygen was in the jar to start with. You should find that the water level rises by about one-fifth of the height of the jar.

These balloons are filled with a gas that is lighter than air. If you let go of their strings, they will float away into heavier air.

MATERIALS

You will need: candle, clear mixing bowl, adhesive, pitcher with colored water, glass jar, felt-tip marker.

Measure the oxygen

1 Secure the candle on the bottom of the bowl with adhesive. Pour colored water into the bowl up to 2 inches (5 centimeters) deep.

2 Ask an adult to light the candle. Then, place the jar over the candle. Let it rest on the bottom of the bowl and watch what happens.

3 The water rises until the candle goes out. Mark the water level on the jar. This will show how much oxygen was present initially.

See the weight

We cannot see the air around us and usually cannot feel it. In fact, we almost forget it is there most of the time. Air seems weightless, but air has weight just like any other material. This experiment shows that a large balloon is heavier than a small one because it contains more air.

MATERIALS

You will need: tape, ruler, thread, 2 balloons of equal size.

1 Place a piece of tape onto the middle of the ruler. Tie a piece of thread around the tape.

2 Lift the ruler and adjust the position of the thread so that the ruler balances.

3 Blow up one balloon a little and the other one much more. Tape them to opposite ends of the ruler.

4 Hold up the ruler with the thread and see what happens. The large balloon makes the ruler dip down. It is heavier than the small balloon because it contains more air.

FEATURES OF WEATHER

Joshua trees in the deserts of southern California. It is very hot there, and hardly any rain falls during the year. Only a few kinds of plants can survive the drought conditions by storing water.

THE three main things that help us describe the weather are temperature, humidity, and pressure. The humidity tells us how much moisture is in the air. Pressure is the pressure of the atmosphere, called atmospheric pressure. The atmosphere has weight and presses down on the ground with a certain force. At sea level, the atmosphere presses down on each square $^1/_2$ inch (3.2 square centimeters) of surface with a force of about 2.2 pounds (1 kilogram). Atmospheric pressure varies a little from place to place across the earth. Differences in pressure cause the air to move around. Air moves from a region of high pressure to one of lower pressure, often bringing about changes in the temperature and humidity.

Hot and wet

This picture *(above)* shows the Everglades in Florida. It's very hot there, but plenty of rain falls during the year. The Everglades National Park is a huge shallow river where hundreds of different plants and animals thrive. Alligators live in the swampy areas.

FACT BOX

- When you hold out your hand, palm upward, you are supporting a weight of about 2.2 pounds (1 kg). This is the weight of the air pressing down on your hand.
- The least humid, or driest, places on earth are the hot deserts. The most humid places are the tropical rain forests near the equator.
- When the weather is hot and humid, our skin feels sticky because, when it is hot, we perspire – our skin gives off tiny drops of water. Normally, the water quickly evaporates, or vanishes, into the air. But, when the weather is humid, the water stays on the skin much longer. That is what makes us feel sticky.

Coldest

The coldest place on earth is Antarctica. The world's lowest temperature was recorded there in July, 1983. It was minus 128.6°F (-89.2°C).

Air pressure

When you pump up the tires on your bicycle, you increase the pressure of the air in them. Increased air pressure makes air masses move around the world.

Hottest

The hottest place on earth is Death Valley in California. Daily summer temperatures may stay above 120°F (49°C) for over a month.

AIR PRESSURE

WHENEVER one thing presses against another, it exerts pressure. Pressure is a kind of force. A heavy book resting on a table exerts pressure on the part of the table underneath it. Air may not be heavy like a book, but it still exerts pressure. Air presses down on everything around us, including human bodies. We cannot feel it because the pressure inside our bodies equals the air pressure outside, so the two cancel each other out. Differences in the pressure of the air in different places causes the air to move around as wind. Here are some tricks that involve air pressure. We can use air pressure to knock over some books. Air pressure also helps show that, sometimes, paper appears to be stronger than wood, because air pressure can hold the paper in place. In a fizzy experiment, other gases, besides air, exert pressure.

MATERIALS

You will need: balloon, books, air pump, strip of wood, newspaper, glove.

Knocking over books
Place a balloon under some books and blow air into it with a pump. As you pump, the air pressure inside the balloon increases, and the force on the books pushes the pile over.

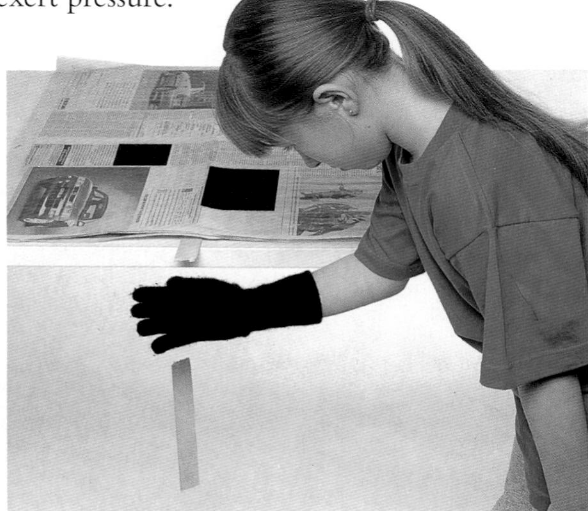

Chopping wood
Lay a strip of wood on a table, letting one-third hang over. Cover it with newspaper. Wear a glove and strike the overhanging wood. It breaks. The paper stays put.

Create air pressure

1 Place the funnel in the neck of the bottle. Pour in some vinegar, filling the bottle about halfway up.

2 Place the funnel in the neck of the balloon. Carefully sprinkle some baking powder into it.

MATERIALS

You will need: funnel, bottle, vinegar, balloon, baking powder.

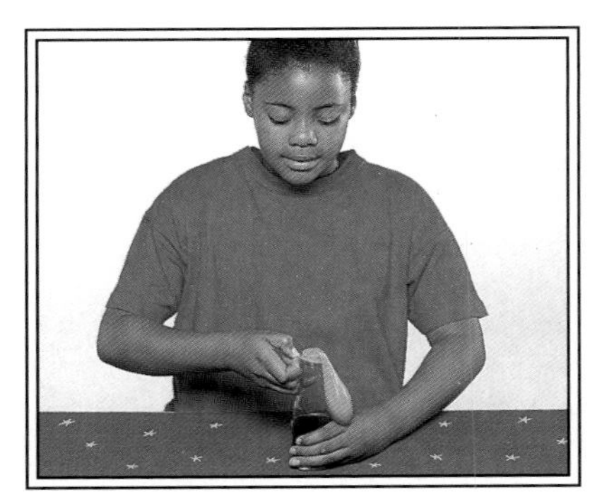

3 Fit the neck of the balloon carefully over the neck of the bottle, letting the powder-filled part hang down. Gradually lift up this part of the balloon.

4 Let the baking powder fall into the vinegar. The balloon will expand as the mixture starts to fizz.

5 The fizzing shows that a chemical reaction is taking place between the vinegar and the baking powder. This reaction makes lots of gas. As more gas is produced, its pressure rises. The rising gas pressure forces the balloon to expand.

AIR ON THE MOVE

Flying a kite is easy, but only when there is a breeze that is not too strong.

We usually do not notice the air until it moves. Moving air is called wind. Winds blow from regions of high pressure to regions of low pressure. Gentle winds are called breezes. These almost always blow at the seaside. They occur because of the difference in temperature between the sea and the land which sets up differences in pressure that cause the breezes. The speed of wind varies greatly. A breeze is a wind with a speed of up to 30 miles (48.3 km) per hour. Gales are stronger winds that travel up to 60 miles (96.5 km) per hour. Storm winds travel at speeds of 75 miles (120.7 km) per hour. Winds stronger than these are called hurricanes. They can cause great damage and destruction. Hurricanes are greatly feared in those parts of the world where they occur.

WIND FORCE

The force, or strength, of the wind varies widely. A scale called the Beaufort Scale is used to describe the force of the wind. The scale goes from Force 0, which means the air is calm, to Force 12, which means a hurricane is blowing. One way you can guess the force of the wind is by the effect it has on you.

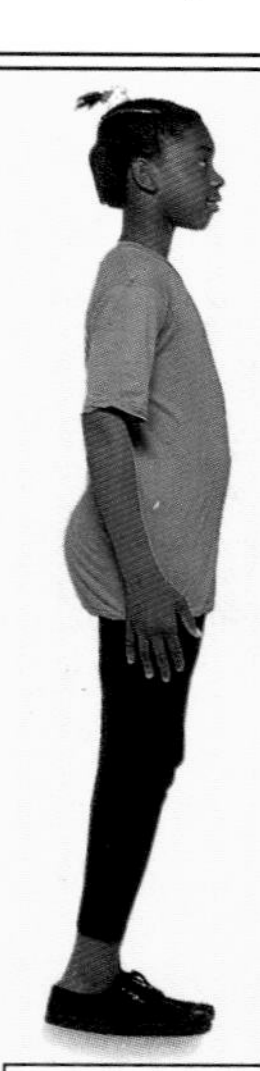

When the wind is Force 0, you cannot feel it. Smoke from chimneys goes straight up. When the wind has reached Force 2, you can feel it on your face, and it is called a breeze.

You can feel a Force 4 breeze pushing against your body when you walk.

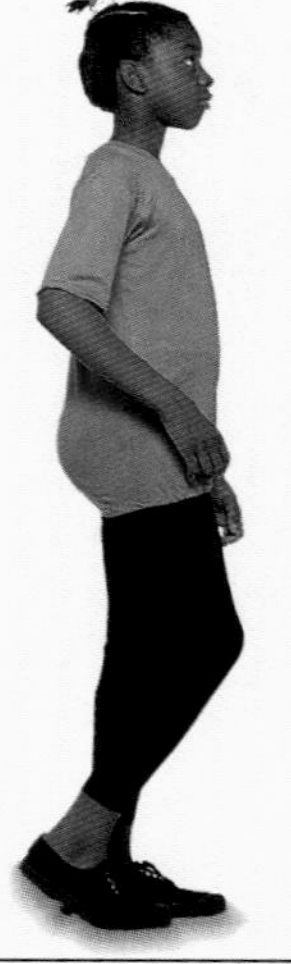

0	1	2	3	4	5

Land breezes and sea breezes
A breeze always blows at the seaside. At night, the land becomes cooler than the sea. Warm air rises above the sea, pulling in cool air from the land as a land breeze. The opposite happens during the day, when a sea breeze blows in from the sea.

Wind power
Giant propellers harness the power of the wind at a wind farm. This one is at Altamont Pass in California. The propellers drive turbines that produce electricity for homes.

At Force 7, the wind has become a gale. You have to bend your body to walk against it.

At Force 9, the wind blows at 50 miles (80 km) per hour or more. You have to squat down, otherwise you will be blown over.

As the wind increases to Forces 10 and 11, you have to stay flat on the ground to keep yourself from being blown away. Do not go out in a Force 12 hurricane – it will take you with it!

MEASURING THE WIND

The harder you blow on your windmills, the faster they spin around.

THE wind plays a very important part in meteorology (the study of weather and climate). It shifts the air from place to place and brings about changes in the weather. Meteorologists measure the direction of the wind to help them forecast where the changes will take place. They also measure the speed of the wind to give them an accurate weather prediction. To measure the wind direction, they use an instrument called a weather vane. Learn how to make a simple one here.

MATERIALS

You will need: adhesive, plastic container (with lid), scissors, twig, 2 plastic straws, colored cardboard, pen, tape, pin, plywood, compass.

Make a weather vane

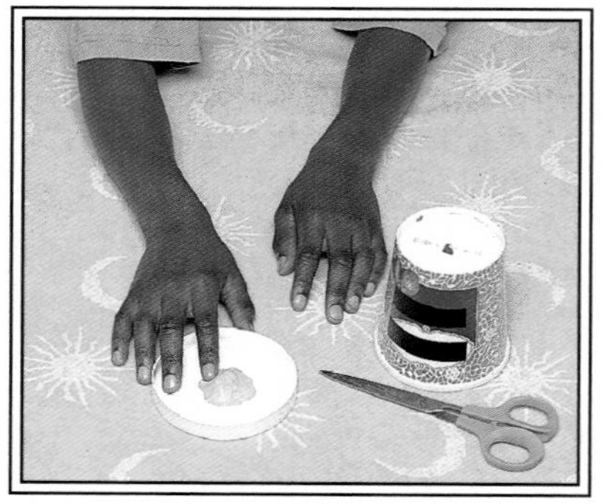

1 Place adhesive in the center of the plastic container's lid. Ask an adult to poke a hole in the container's bottom with the scissors. Put the lid on the container.

2 Slide a piece of twig into a straw. Trim the twig so that it is a bit shorter than the straw. Push the straw and the twig through the container's hole into the adhesive.

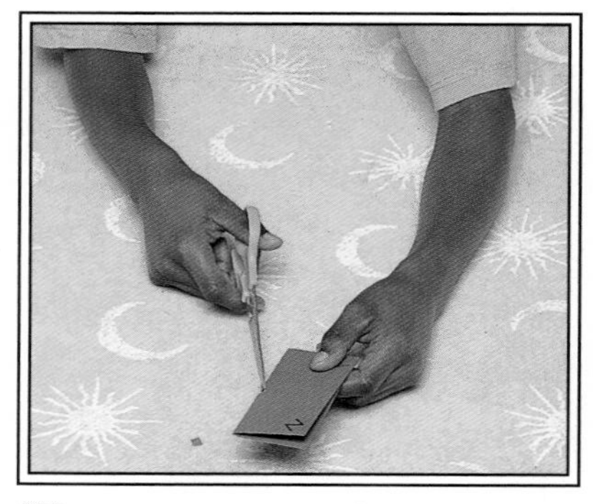

3 Cut out a square of cardboard and mark each corner with a point of the compass – N, S, E, W. Snip a hole in the center and carefully slip the cardboard over the straw.

In a spin
Your windmill will give you an idea of how fast the wind is blowing. The faster the wind blows, the faster the windmill will spin.

Windy ways
When you have made your weather vane, take it outside. Use a compass to turn your cardboard compass so the corners point in the right directions. Now, let the wind blow!

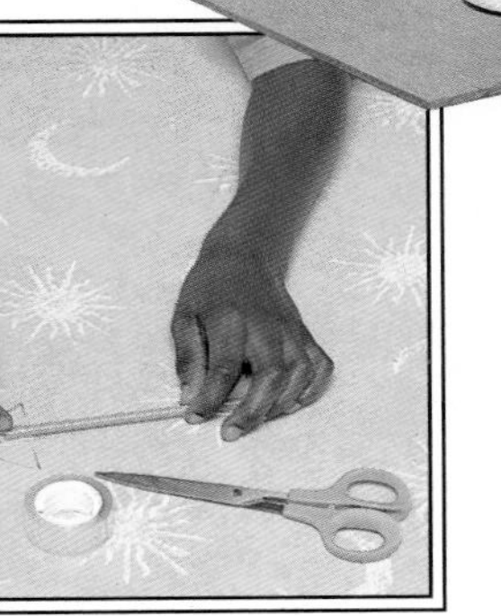

4 Cut out two cardboard triangles. Tape them to each end of another straw to form the head and tail of an arrow. Put some adhesive in the top of the first straw.

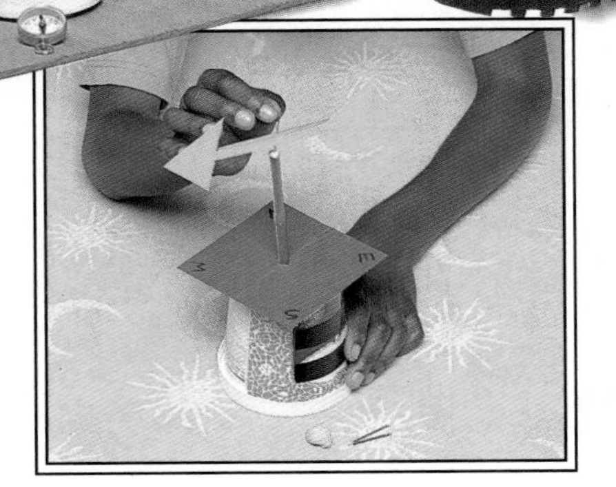

5 Push a pin through the middle of the arrow straw and stick it into the adhesive in the first straw. Be careful when using sharp objects like pins.

6 Secure your weather vane on plywood with adhesive. Test it for use. The arrow should spin around freely when the wind blows on it.

STORMY WEATHER

A spiraling column of wind picks up the dust, creating a dust devil. Dust devils often form over hot, dry farmland in summer.

WHEN the weather is very rough and windy, we say it is stormy. But, strictly speaking, a *storm* is the name for a particular kind of strong wind. Sometimes, on hot, dusty days in summer, little dust storms spring up. They are called dust devils. These little whirling winds pick up the dust but do not do any harm. Some whirlwinds, however, such as tornadoes, are very destructive. They are great funnels of wind, in which the wind rushes around at speeds of up to 300 miles (483 km) per hour. They form over the land. Hurricanes, which form over the sea, are much larger. They can measure more than 300 miles (483 km) across.

Spiral of wind
This is what a hurricane looks like from space. The clouds form in a great spiral, getting thicker toward the center. On the ground, the winds reach speeds approaching 125 miles (200 km) per hour.

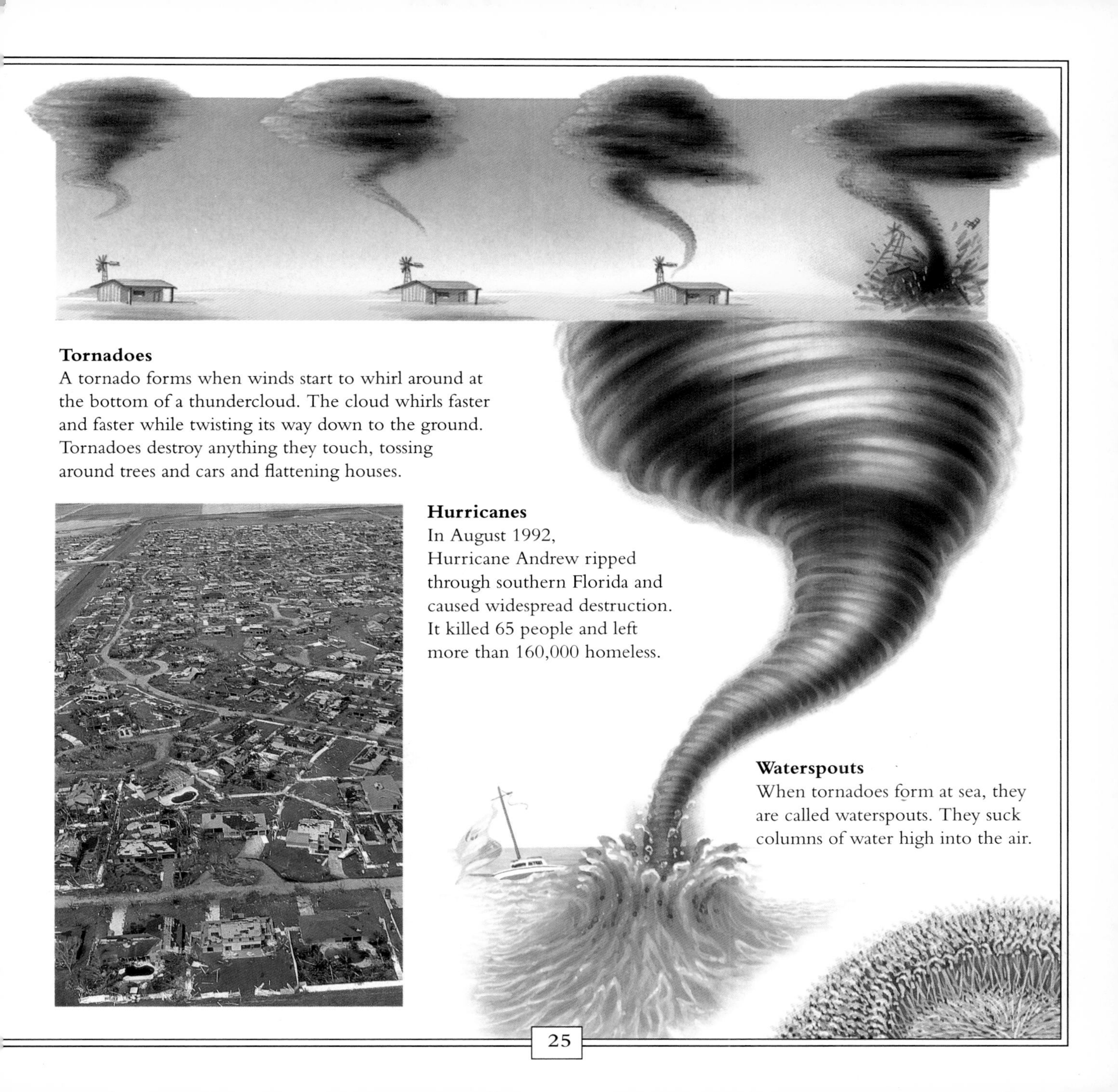

Tornadoes

A tornado forms when winds start to whirl around at the bottom of a thundercloud. The cloud whirls faster and faster while twisting its way down to the ground. Tornadoes destroy anything they touch, tossing around trees and cars and flattening houses.

Hurricanes

In August 1992, Hurricane Andrew ripped through southern Florida and caused widespread destruction. It killed 65 people and left more than 160,000 homeless.

Waterspouts

When tornadoes form at sea, they are called waterspouts. They suck columns of water high into the air.

THE WATER CYCLE

THE amount of moisture, or water, in the air has a great effect on the weather. For example, if there is a lot of water in the air, the chances are it will rain. On the other hand, if the air is dry, the weather should be clear. The amount of moisture in the air is called humidity. Water usually is found in the air as a gas called water vapor. The water vapor gets into the air mainly from the oceans, lakes, and rivers. It is produced when the heat from the sun warms up surface water and makes it evaporate, or turn to gas, called water vapor. The water vapor rises into the air. The air is cooler higher up. When water vapor cools enough, it condenses, or changes back into tiny droplets of liquid water. The droplets gather together to form the great billowing masses we call clouds. Often, the droplets grow bigger and bigger until they become heavy enough to fall from the clouds, as rain – or snow, if it is cold enough.

This movement of water from the ground to the air and back again occurs continually. It is called the water cycle.

Water vapor turns into droplets and forms clouds

FACT BOX

- Most water vapor in the air comes from the oceans, which cover more than 70 percent of the earth. Water vapor is created when water from the ocean surface evaporates.
- Each day, about 30,500 cubic miles (127,125 cubic kilometers) of water evaporate from the oceans and turn into water vapor. About the same amount of water returns to the earth each day, mainly in the form of rain and snow.
- It takes water evaporated from the oceans about nine days to return to the earth's surface.

The sun provides the energy that powers the water cycle. It pours heat onto the earth, which makes water evaporate from the rivers and the seas. The water vapor rises into the sky, where it cools and turns into droplets of liquid water.

Water evaporates from sea

In a cloud

High up on Mount Tasman in New Zealand, you might find yourself in a cloud, because, higher up a mountain, the air becomes cooler. When it is cool enough, clouds form. You can feel how damp the air becomes.

HUMIDITY

The forests along the coast of northern California stay warm and humid most of the time. In this climate, the redwood trees there can grow to more than 100 yards (90 meters) tall.

THERE is always a certain amount of water in the air, in the form of a gas called water vapor. Meteorologists call the amount of water vapor in the air the humidity. When there is a lot of water vapor around, the air feels humid, or sticky. When there is little water vapor around, the air feels dry. Measuring the amount of water vapor in the air helps meteorologists with their forecasts. When the air gets very humid, for example, the chances are that it soon will rain. An instrument called a hygrometer is used to measure humidity. Learn how to make a simple one here.

MATERIALS

You will need: 2 sheets of colored cardboard, scissors, pen, glue, toothpick, used matchstick, straw, adhesive, blotting paper, hole punch.

Measure the humidity

1 Cut out a cardboard rectangle. Draw marks along one side, at regular intervals, for a scale. Cut a slit 1 inch (2.5 cm) long in the bottom. Fold and glue these flaps to a cardboard base, as shown *above*.

2 Cut another longer cardboard rectangle. Fold and glue it to the base, as shown above. Pierce the top of the two folds carefully with a toothpick to create a pivot.

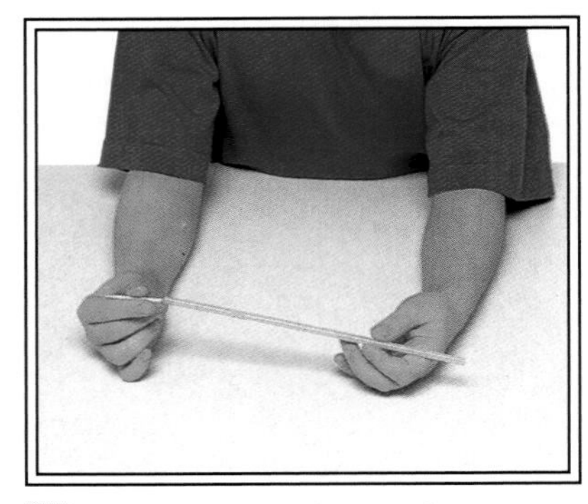

3 Make a pointer by attaching a used matchstick to one end of a straw with adhesive. The matchstick gives the pointer some weight and keeps it from easily being knocked off balance.

Transpiring plants

Plants play an important part in the water cycle because their leaves give off water vapor. This is called transpiration. Cover a potted plant with a clear plastic bag, sealing it around the pot with tape. Place the plant in the sunlight for two hours. Notice that the bag starts to mist up. Droplets of water form inside the bag when the water vapor given off by the plant condenses, or turns back to liquid.

4 Cut several squares of blotting paper and punch a hole in the center of each one. Slide the squares over the end of the pointer.

5 Pierce the pointer with the toothpick pivot and position the pointer as shown above. Make sure it can swing freely up and down.

6 Your hygrometer is now ready for use. Adjust the position of the toothpick so that the pointer is horizontal. Take your hygrometer into the bathroom when you take a bath or shower. The high humidity in the bathroom should make the blotting paper damp. It will tip the pointer up. Outside, on a warm day, the blotting paper will dry, and the pointer will tip down.

LOOKING AT CLOUDS

CLOUDS are great fluffy masses of tiny water droplets or ice crystals hanging in the sky. The kinds of clouds we see reveal much about the weather. There are three main kinds of clouds. The most familiar kind is the cumulus, the fluffy cloud of fine summer days that looks like a ball of cotton. A stratus cloud, on the other hand, is flat and can stretch all across the sky. It brings dull weather and, often, rain. The stratus is a low cloud. The third main kind of cloud – cirrus – is a very high cloud that can form up to 8 miles (13 km) up in the air. The cirrus is a wispy, feathery kind of cloud made up of ice crystals rather than water droplets. Dark clouds that bring rain are called nimbus. Cloud names can be combined. For example, a cumulonimbus cloud is a dark cumulus cloud that brings heavy rain.

Clouds form when warm air containing water vapor rises and cools. The water vapor turns into droplets of water, forming clouds. If the air is very cold, the water vapor turns into a cloud of tiny ice crystals.

Cumulus
These unusual cumulus clouds look a lot like teeth. They are lit dramatically by a low evening sun.

Cirrus
A fine cirrus cloud formation. You can see why this type of cloud is often called mares' tails.

Cumulonimbus
This great cumulus cloud is expanding into a thunderclound and will soon develop into the anvil shape of a cumulonimbus.

Cirrocumulus
A classic mackerel sky, so-called because it looks a lot like the pattern on the back of the mackerel fish. These clouds are called cirrocumulus.

The main kinds of clouds

RAIN AND DEW

In some clouds, the water droplets remain tiny and stay up in the air. In others, the droplets keep bumping into one another and joining together. They get larger and larger, eventually becoming heavy enough to fall out of the cloud as rain. On average, raindrops measure only about 1/10 inch (2.5 millimeters) across, but the drops that fall from thunderclouds are much bigger. Rain is the most common form of what meteorologists call precipitation, which is a deposit that comes out of the air and falls to the ground. Dew is another kind of precipitation. It forms on the ground and other surfaces on cool nights. The cool surfaces make the water vapor in the air condense into liquid droplets of water.

It can be fun going out in the rain, if you are dressed properly. But watch that umbrella if the wind is strong. It may turn inside out or fly away!

Rain to come
Stormy weather off the island of Majorca in the Mediterranean Sea. The dark nimbus clouds pile up. The sun seems to be drawing water from the sea. Soon it will be raining hard.

Heavy monsoon rains occur in the summer in India when very warm, moist winds blow in from the Indian Ocean.

Rainbows
When it is raining and the sun is low in the sky, you often can see a rainbow. This beautiful band of colors is produced when sunlight passes through raindrops.

Lots of rain
The monsoon rain falls in Jaipur, India. In a few months, 30 feet (9 m) of rain may fall. The rest of the year is parched and dry.

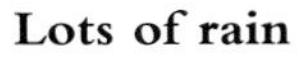

Moist air
In hot, tropical regions, such as Hawaii, rain falls regularly throughout the year. The heat and humidity, or moisture, provide perfect growing conditions for all kinds of plants.

Dew drops glisten on the delicate strands of a spider's web in the early morning.

MAKING RAINBOWS

RAINBOWS often appear in rainy weather, when the sun is quite low in the sky. They form because raindrops split up the light from the sun into a spread of different colors. This spread is called a spectrum. In science, white light can be split up into a colorful spectrum by shining it through a prism, or a wedge of glass. In this experiment, a spectrum is produced by shining light through a wedge of water.

Seven main colors can be seen in the rainbow. They are red (on the outside), orange, yellow, green, blue, indigo, and violet (on the inside).

MATERIALS

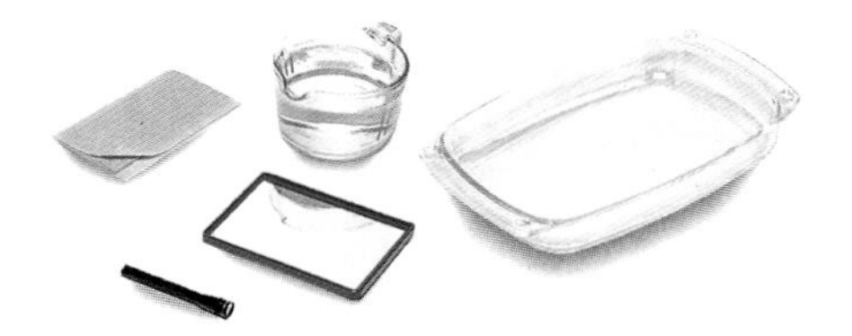

You will need: mirror, oblong pan or dish, adhesive, pitcher of water, flashlight, white cardboard.

Split light into a rainbow

1 Lean the mirror carefully against the side of the pan. Secure it with adhesive so that it slopes at an angle.

2 Pour water into the pan until it is 1-2 inches (2.5-5 cm) deep, creating a wedge of water by the mirror.

3 Turn on the flashlight. Shine the beam onto the surface of the water in front of the mirror to produce a spectrum, or rainbow.

FACT BOX

• White light is not really white. It actually is made up of light of many colors that combine to make white.

• When light travels through the air into water or glass (or back the other way), it bends. Different colors in the light bend more than others. Blue light bends the most, and red light the least, causing the colors to separate. The result is a spectrum – the colors of the rainbow.

This is what happens in a rainbow.

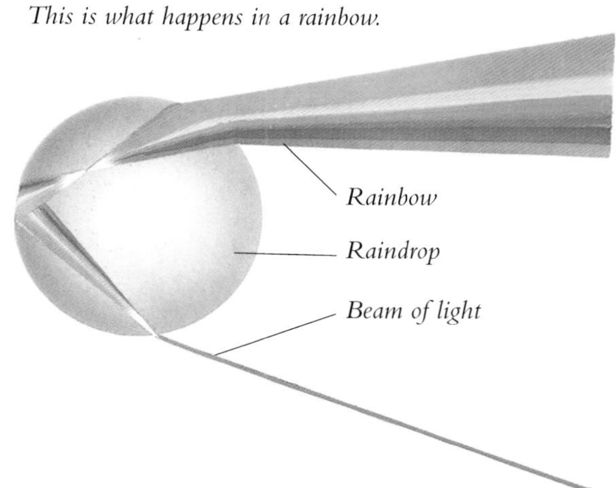

This is what happens in your experiment.

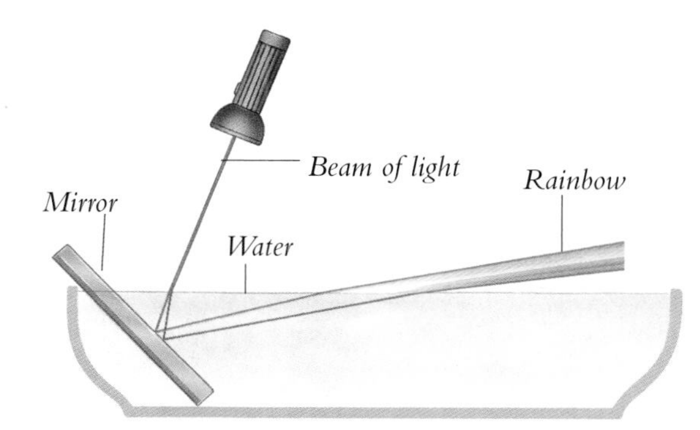

4 To look at your rainbow, hold the white cardboard above the dish. You may need to alter the positions of the cardboard and the flashlight before you can see the rainbow properly. It is best if you do this part of the experiment in dim light.

THUNDER AND LIGHTNING

Sieou-wen-ing sends bolts of lightning toward the earth. In Chinese mythology, she is the mother of lightning.

A thunderstorm is one of nature's greatest spectacles. Huge, dark thunderclouds, as much as 6 miles (10 km) deep, tower up through the sky. Meteorologists call them cumulonimbus. Lightning flashes zigzag between the clouds and fork down to the ground. The air is filled with cracks and rumbles of thunder. But what exactly are lightning and thunder? Lightning is a giant electric spark. Electricity builds up in thunderclouds as the drops of water and lumps of ice whirl around. Eventually, the electricity becomes so powerful that it jumps to other clouds or to the ground. The air in its path gets very hot and expands suddenly, creating a kind of explosion, which we hear as thunder.

Struck by lightning
Lightning has struck this tree *(right)*, leaving a trail of exposed wood on the bark.

Hail
Lumps of ice, called hailstones, often fall out of thunderclouds. Here, hailstones have covered a street in Pretoria, South Africa.

WARNING

If you are caught outside in a thunderstorm, do not take shelter under a tree. The tree might be struck by lightning, and the lightning could jump from the tree to your body. This could knock you out, or even kill you.

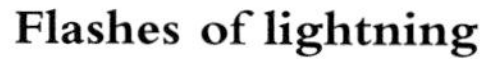

Flashes of lightning
A spectacular night display of lightning above distant mountains. Note that the lightning not only is traveling to earth but also is jumping around between the clouds.

Lightning conductor
This church steeple has a lightning conductor. If lightning should strike, it will travel safely down the lightning conductor without causing damage to the rest of the building.

PROJECT

GAUGING THE RAIN

In most parts of the world, it rains on and off throughout the year. In some places, more than several feet of rain can fall in a year. In others, there is much less rain. How much rainfall do you get where you live? Make this simple rain gauge and find out. Gauge is a name for a measuring instrument. If it is very rainy, this project will keep you busy. If you live in a desert area, perhaps you had better pass on this experiment. You might have to wait for years to try out your rain gauge!

Don't forget your umbrella and rubber boots in the rain.

MATERIALS

You will need: scissors, tape, large wide jar, ruler, ballpoint pen, large plastic funnel, notebook, tall narrow jar or bottle.

Measure rainfall

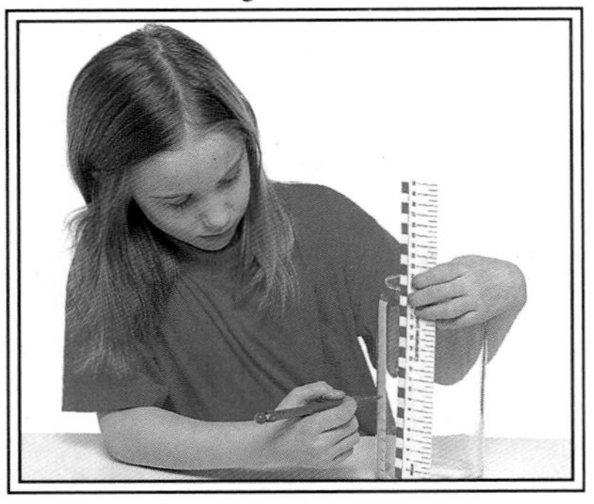

1 Cut a length of tape the same height as the large jar and stick it along the side of the jar. Using a ruler, mark a scale on the tape at ½-inch (1-cm) intervals. Measure the diameter of the jar and cut the funnel to exactly the same size.

2 Place the funnel in the jar, and your rain gauge is ready. Put the gauge outside in an open space away from trees or shrubs. Look at your rain gauge at the same time each morning or evening to see if it rained in the last 24 hours.

3 When it has rained, look at the scale to see how much water is in the jar. This is the rainfall for the past 24 hours. Write down the amount in your notebook. Be sure to empty the jar before you put it back outside for the next reading.

Being precise

You can measure the rainfall more accurately if you use a separate narrow measuring jar or bottle. First, stick another strip of tape along the side of this bottle. Pour water into the large collecting jar up to the $^1/_2$-inch (1-cm) mark. Now pour this water into the measuring bottle. Mark $^1/_2$ inch (1 cm) where the water level reaches. Divide the length from the bottom of the bottle to the $^1/_2$-inch (1-cm) mark into ten equal parts. Each part will be equivalent to .05 inch (1 mm) of rainfall. Extend the scale past the $^1/_2$-inch (1-cm) mark to the top of the measuring bottle. Ask an adult to help you with this part of the experiment. You can use this bottle to measure the rainfall you collect, accurately, to the nearest hundredth of an inch ($^1/_4$ mm), just like professional meteorologists do.

Looking at rain

Some companies make all-in-one weather instruments, which are very handy when space is limited. With this particular instrument you can measure the temperature, amount of rainfall, wind direction, and wind speed. This girl is reading the rain gauge on her all-in-one instrument.

FACT BOX

- The wettest place in the world is Mawsynram, in India, where an average of nearly 40 feet (12 m) of rain falls every year.
- New York and Sydney have a little over 40 inches (102 cm) of rain a year. Paris and London have about 24 inches (61 cm) of rain a year.
- The driest place in the world is the Atacama Desert, west of the Andes in Chile, South America. Just a few showers of rain fall in parts of the desert every century.
- Sea storms can cause worse flooding than rainfall. Large waves form that submerge coastal areas.

SNOW AND ICE

The beautiful shape of a snowflake, formed from delicate strands of ice. Snowflakes are among nature's greatest works of art.

SNOW falls in winter in many countries. It also falls all year round in places near the North and South poles and at the top of high mountains. Snow is a form of precipitation that falls to the ground from the clouds. It forms at the tops of high clouds where the temperature is below freezing. Clusters of snowflakes fall from the clouds when they become heavy. But, if the lower air is warm, they melt and fall as rain. Sometimes, a mixture of snow and rain falls as sleet. A snowflake is a mass of tiny crystals of frozen water, or ice. If you look at a snowflake under a microscope, you will see that it has the shape of a six-pointed star. On many winter nights, the ground becomes snow-white even when it has not been snowing. This white covering is frost. Frost forms when the ground gets cold and water vapor in the air condenses on it. The water immediately freezes into tiny sparkling crystals of ice.

Cover of snow
Thick snow has fallen in the Austrian Alps. Snow falls all through winter in the Alps, the peaks of which rise to more than 15,000 feet (4,572 m).

Ice on glass
Look at the frost on a windowpane. The ice has formed feathery crystals where it froze.

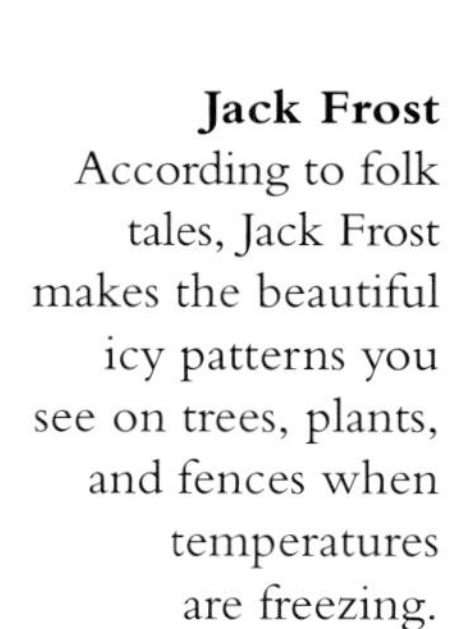

Jack Frost
According to folk tales, Jack Frost makes the beautiful icy patterns you see on trees, plants, and fences when temperatures are freezing.

Chunks of ice
Most of the ice in the world is found in the ice caps at the North and South poles. Great chunks of ice constantly break off and float away as icebergs.

MASSES OF AIR

Air masses cross paths, and a storm may occur. After a storm, dark clouds still hang overhead near Krissavik, Iceland.

GREAT bodies of air are moving through the atmosphere all the time. Each has a different temperature and contains a different amount of moisture. They are called air masses. While a single air mass is passing us by, the weather remains the same. When another air mass comes along, the weather changes. When two air masses come up against each other, the weather almost always takes a turn for the worse. Thick clouds may form, and storms break out at the boundary, or front, where the air masses meet. The weather settles down again after the front passes.

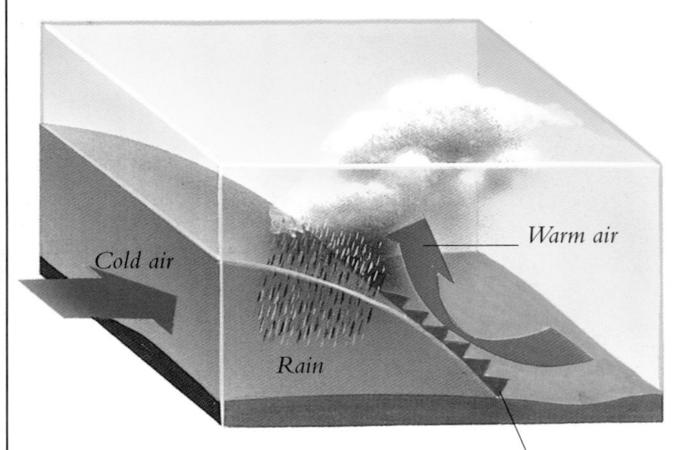

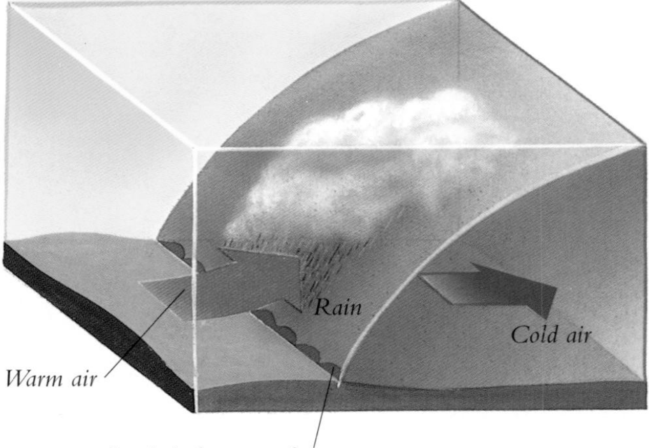

Cold front
Clouds form on a cold front as a wedge of cool air pushes underneath a mass of warm air. The warm air is forced to rise. As it cools, clouds form and rain falls.

Warm front
When a warm front moves in, the warm air rides up over the cold air, and clouds form. Usually rain falls, as shown in the drawing *above*.

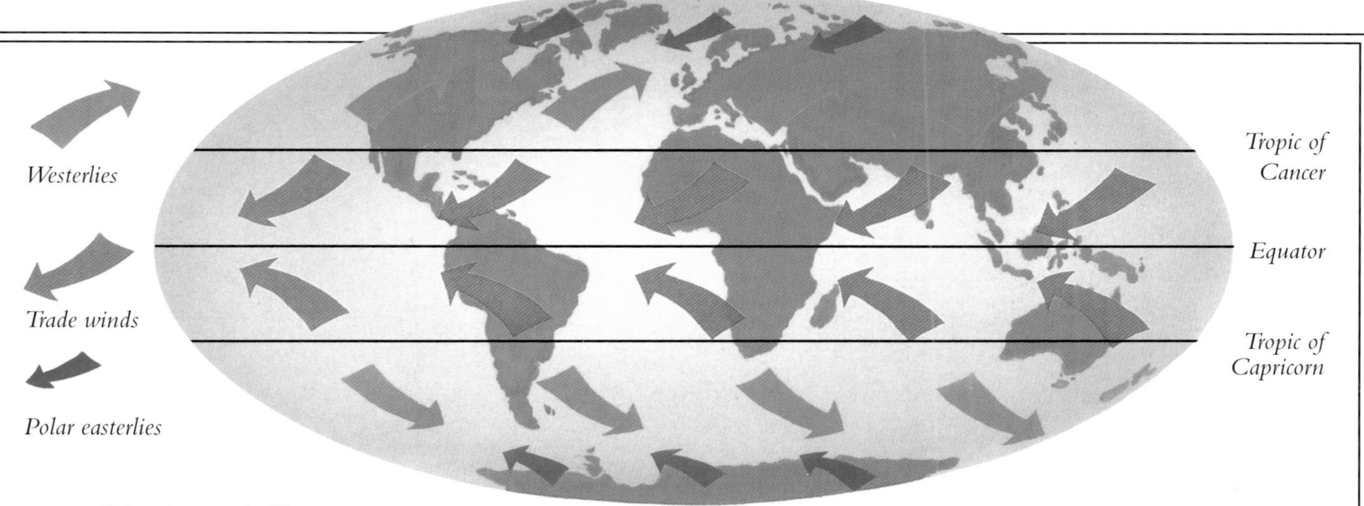

World winds

This map *(above)* shows the major air movements around the world, called wind belts. They are prevailing winds, which nearly always blow in the same direction. The trade winds are warm because they blow on either side of the equator. The westerlies are cool and blow north of the Tropic of Cancer and south of the Tropic of Capricorn. The polar easterlies are icy and blow around the North and South poles.

Polar fronts

Great masses of warm and cool air move around the earth. Warm air masses travel north and south away from the equator. Cold air masses travel south and north away from the North and South poles. They meet along boundaries called polar fronts.

Fast winds

On the open sea, sailing boats and ships have always relied on the trade winds and westerlies to speed them around the world.

HOLDING THE HEAT

MOST climates vary around the world because different places receive different amounts of the sun's heat. However, other things affect the climate, too. Places on seacoasts have a different climate from places inland, because land and water hold onto heat differently. Water takes longer than land to warm up and cool down. So, summers are cooler and winters are milder on the coast than they are inland. Try this experiment with sand (land) and water (sea).

MATERIALS

You will need: pitcher of water, sand, 2 bowls, thermometer, watch, notebook, pen.

When you put sand and water in the sunshine, you can feel how much warmer they become.

Measure temperature changes

1 Pour water into one bowl and put sand in the other. You do not need to measure them, but use approximately equal amounts.

2 Put the bowls, side by side, in a cool place indoors. After a few hours, record the temperatures of their contents. Are they the same?

3 Place the bowls, side by side, in the sunlight outside for an hour or two. Record the temperatures of the sand and water again.

Dark and light

In this experiment, you will find that dark things and light things heat up and cool down differently. In hot countries, such as Saudi Arabia, wearing white robes keeps people cooler because white things do not heat up as much as dark things.

MATERIALS

You will need: 2 identical glass jars with lids (one painted black, one painted white), sand, spoon, watch, thermometer, notebook, pen.

1 Fill the two jars with sand to about the same level. Screw the lids on firmly. Place both jars outside in the sunshine and leave them there for two hours.

2 Take the temperature of the sand in each jar. The sand in the jar with the black lid should be hotter than the sand in the other jar. Write these temperatures in a notebook.

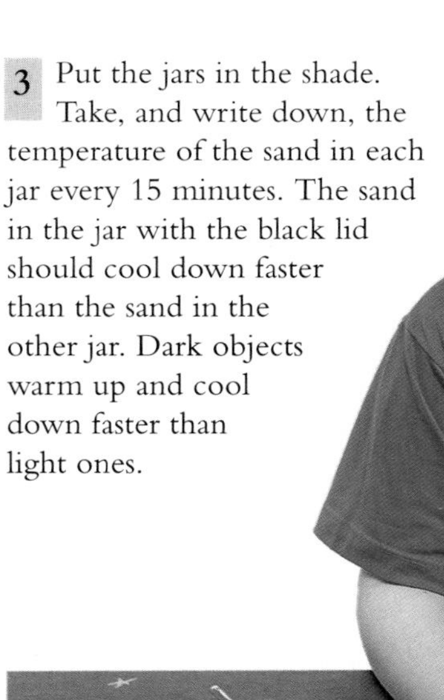

3 Put the jars in the shade. Take, and write down, the temperature of the sand in each jar every 15 minutes. The sand in the jar with the black lid should cool down faster than the sand in the other jar. Dark objects warm up and cool down faster than light ones.

4 Take the bowls back indoors to a cool place. Record the temperatures of the sand and water every 15 minutes. Sand will cool down faster than water.

WORLD WEATHER

DIFFERENT parts of the world receive varying amounts of heat from the sun. So, different places have different climates. Places near the equator have the hottest climates. Climates gradually get cooler as you move north and south away from the equator. The world can be divided into regions with similar kinds of climates. These regions are shown on the map *(right)*. In tropical regions, it is hot and wet. In dry desert regions, it usually is hot, and scarcely any rain falls. In temperate regions, it is not too hot and not too cold, and there is a reasonable amount of rainfall. Mountain regions have a climate that changes with height. In the polar climate, near the North and South poles, the weather is very cold almost all of the time.

This map shows the main kinds of climates found throughout the world. Different kinds of animals and plants live in each kind of climate.

Key to climates

Polar
Mountain
Cold forest

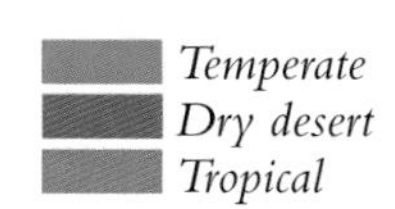

Temperate
Dry desert
Tropical

The cotton-top tamarin lives in the rain forests of Colombia in South America. It thrives in a hot, humid climate.

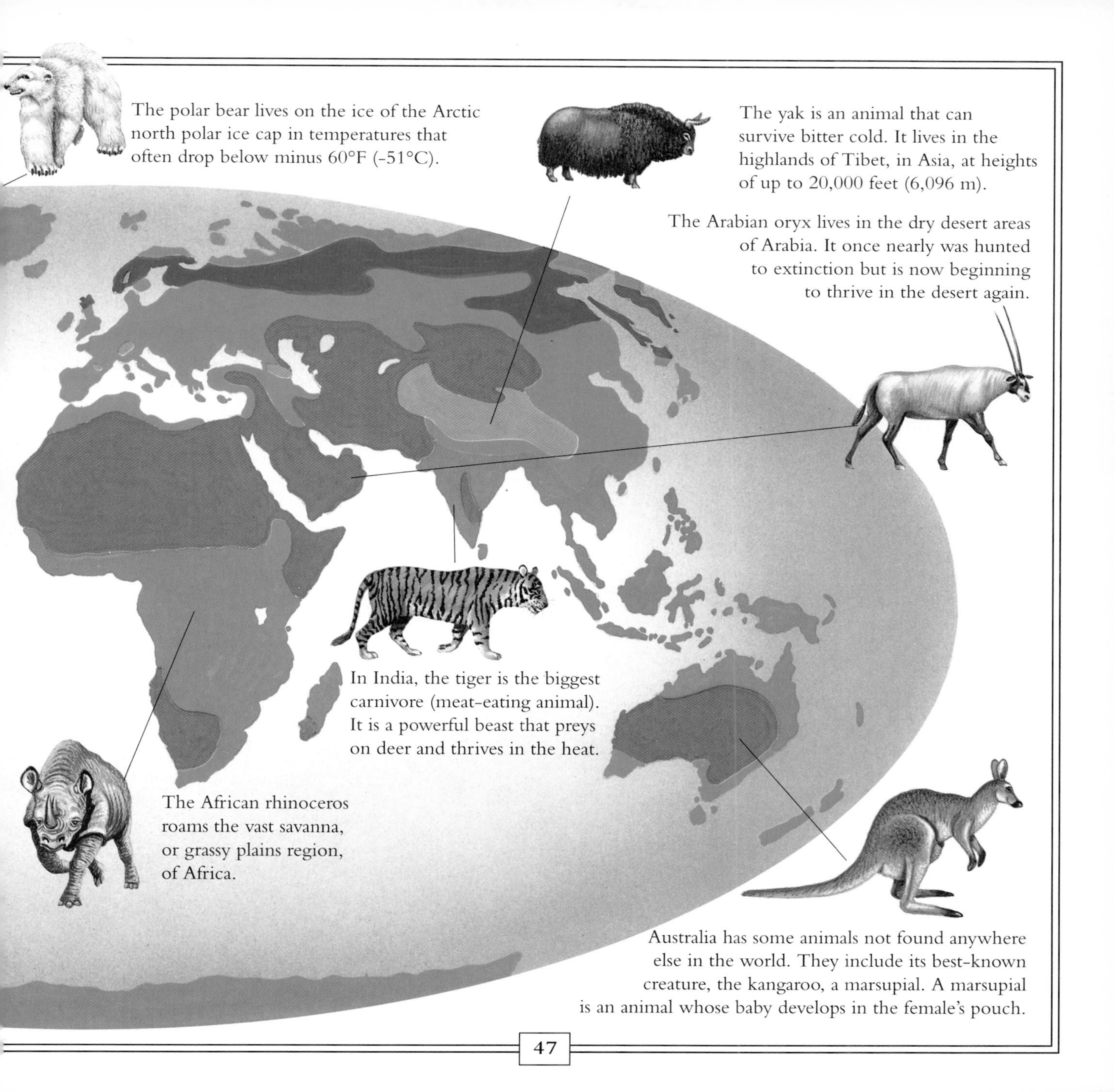

The polar bear lives on the ice of the Arctic north polar ice cap in temperatures that often drop below minus 60°F (-51°C).

The yak is an animal that can survive bitter cold. It lives in the highlands of Tibet, in Asia, at heights of up to 20,000 feet (6,096 m).

The Arabian oryx lives in the dry desert areas of Arabia. It once nearly was hunted to extinction but is now beginning to thrive in the desert again.

In India, the tiger is the biggest carnivore (meat-eating animal). It is a powerful beast that preys on deer and thrives in the heat.

The African rhinoceros roams the vast savanna, or grassy plains region, of Africa.

Australia has some animals not found anywhere else in the world. They include its best-known creature, the kangaroo, a marsupial. A marsupial is an animal whose baby develops in the female's pouch.

WARM CLIMATES

One of the best-known desert plants in the southwestern United States is the saguaro cactus, or the organ pipe cactus (above). *Cacti are specially adapted to live in drought conditions, because they can store water.*

In some regions near the equator, the climate stays hot all year round, and plenty of rain falls. With all this heat and moisture, plants grow quickly and keep growing all year long. Great forests flourish in these regions. They are called rain forests because of heavy rainfall. Thousands of different species, or kinds, of animals live there and can find plenty to eat all the time. Some areas in equatorial regions become almost completely dry, forming deserts. In other areas, heavy rain may fall for only some of the year, which is what occurs in the African grasslands known as the savanna. Farther from the equator, there are regions with a temperate, or mild, warm climate. Most of Europe and the United States enjoy this kind of climate. Much of the world's most productive farmland is found in these regions.

In the desert

A Bedouin gazes across the largest desert in the world, the Sahara Desert in northern Africa. The Sahara covers an area that is nearly as big as the United States. *Sahara* is the Arabic word for "desert." Most of the Sahara consists of hot sand dunes like those pictured *(right)*. Other areas, however, are hard and rocky.

Temperate plants
In temperate climates, trees, such as beeches, grow very well. They are deciduous, which means they shed their leaves when it gets cold in the autumn.

Tropical plants
Trees grow and flower all through the year in the hot, damp tropical forests. Plant life in these forests grows rapidly, forming dense jungle.

African savanna
Scattered trees grow on the African savanna, where the climate is warm throughout the year. Rain falls only during the wet season; the rest of the year is dry.

Tropical animals
Alligators enjoy lying in the sun in the hot, moist climate of Florida, but they can run as fast as you!

COOL CLIMATES

High in the mountains, the climate is always cool. The main kind of tree is the evergreen conifer. Conifers do not shed their leaves, or needles, in the autumn. They keep them all through the year.

THE northern parts of North America, Europe, and Asia have a cold, temperate climate. The winters are long and cold, and plenty of snow falls. The main plant life of this region is evergreen forest. It forms one of the two largest forest regions still remaining on earth. The tropical rain forests form the other big forest region. At the northern tips of North America, Europe, and Asia, it is too cold for trees to grow. These regions are called the tundra. In the tundra, the winters are very long and harsh. Nevertheless, some plant life manages to survive, taking advantage of the few weeks of summer sun. North of the tundra, around the North Pole, the ground is permanently covered with ice. Temperatures drop below minus 140°F (-95°C) in the long, dark winters. At the other end of the earth, the region around the South Pole has a similarly severe climate.

Hardy reindeer
Reindeer are among the hardiest of animals. They live in the cold, northern forests of the world and on the icy tundra of the Arctic. They grow heavy coats for the winter. Their hoofs are broad and flat to stop them from sinking too far into the snow. In winter, they survive by eating plants called lichens.

Well-insulated seal
Many kinds of seals live in the cold waters of the Arctic. A seal's body is covered with a thick layer of fatty blubber, which keeps out the cold.

Aquatic beaver
The beaver spends much of its time in the water, too. It has a waterproof furry coat. Beavers are found widely in the cool northern forests of North America.

Mount Kenya
A high mountain has several different climates on its slopes because the temperature falls with higher and higher altitudes. Here *(left)* are the main kinds of climates found on Mount Kenya, which is located near the equator.

Sure-footed goat
A mountain goat straddles a rock cleft. It likes the cool temperatures of the mountains and is agile enough to clamber up the steepest slopes.

SEASONAL WEATHER

Ancient Britons built Stonehenge on Salisbury Plain as a means of marking the passing of the seasons.

In most parts of the world, the day-to-day weather changes gradually throughout the year with the seasons. In the United States, Canada, and Europe, which are in the Northern Hemisphere, it is cold in December (winter), warmer in March (spring), hot in June (summer), and cooler in September (autumn). The seasons are reversed in the Southern Hemisphere. The temperature changes with the seasons because of the way the earth is tilted in space as it travels around the sun. However, some parts of the world do not have these four seasons. For example, near the equator, the weather stays hot and wet throughout the year. On the African savanna, there are just two seasons – dry and wet.

North for the summer
In the Northern Hemisphere in summer, swallows build their nests and raise their young.

South for the winter
In the Northern Hemisphere in autumn, swallows gather on telephone lines and prepare to fly south for the winter.

Seasons

The seasons occur because of the way the earth spins as it circles in space around the sun. The axis of the earth is tilted at an angle to the direction it is traveling. So certain places on earth are tilted toward and then away from the sun as the year passes. The more a place is tilted toward the sun, the warmer it is. The more it is tilted away from the sun, the cooler it is.

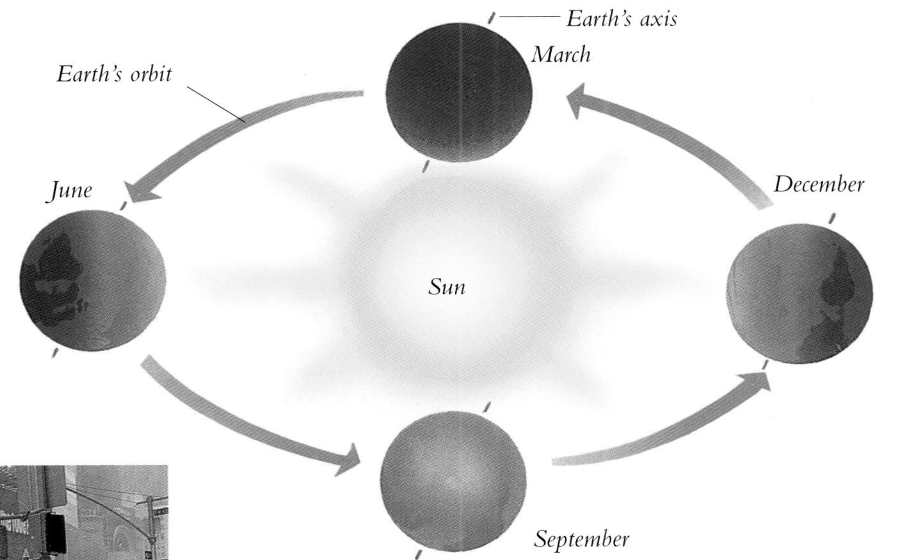

Snowy December

A snowy scene in New York City around Christmas time. It is near mid-winter. The temperature is only just above freezing, and there is a biting wind.

Sunny December

Many thousands of miles away, in Australia, it is mid-summer. Temperatures are about 85°F (29°C). Australia is in the Southern Hemisphere where the seasons are opposite those in the Northern Hemisphere.

CHARGING UP

IN many parts of the world in summer, the weather can get very hot and sticky. Then, we know that thunderstorms are likely to happen, producing lightning flashes and thunder.

Lightning occurs when electricity builds up in the thunderclouds. It is a kind of electricity called static electricity, and it is made up of tiny bits of electricity called electric charges. Lightning is produced when very high electric charges build up in clouds. Tiny electric charges can be built up using balloons.

Standing on end

If you brush your hair when it is very dry, you might make it electrically charged. You can make it stand on end, and you might even hear crackling noises as tiny sparks fly around in your hair.

Make static electricity

1 Blow up a number of balloons. Rub them against a sweater or something made of wool. Put the balloons in different places.

2 Put them on the ceiling, on the walls, on your friends! They stay in place because of electricity that does not move – static electricity.

MATERIALS

You will need: hairbrush, balloons, wool sweater.

MATERIALS

You will need: plastic tablecloth, tape, rubber gloves, metal dish, fork.

Lightning zigzags across the sky as electric charges build up in the clouds and jump around.

Jumping electricity

In the last experiment, you built up tiny electrical charges on balloons by rubbing. The balloons stuck to things because of the attraction of these charges. Up in the clouds, high electrical charges jump around. They create the flashes called lightning. You can make electricity jump with this experiment.

1 Lay out the plastic tablecloth and tape it to the table to prevent it from sliding around and disrupting your experiment.

2 Wearing a rubber glove, slide the metal dish around over the tablecloth for a few minutes. The dish will charge up with electricity.

3 With your ungloved hand, bring the fork close to the dish. You should see a spark jump. It is easier to see in the dark.

THE CHANGING CLIMATE

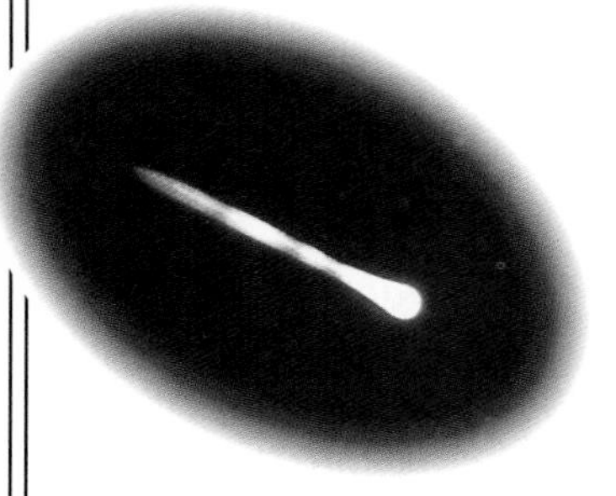

A fireball streaks through the night sky. It is a lump of rock burning up as it plunges down to earth as a meteorite. Big meteorites can bring about a change in climate.

THE weather in a region may change from day to day, but the climate, or the overall weather pattern, remains much the same. Over long periods of time, however, the climate does change. Changes in the heat the sun gives out can make the earth hotter or colder. It causes ice ages, for example, when vast sheets of ice grow to cover large areas of the earth. Big meteorites can cause sudden changes in climate when they hit the earth and kick up huge amounts of dust into the atmosphere. The dust stops the sun's heat from reaching the ground, causing the temperature to fall steeply. Gradual changes are happening because of humans burning fuels. This releases carbon dioxide gas into the atmosphere and turns the earth into a kind of greenhouse, slowly heating our world through global warming.

Trees and climate
Scientists study past changes in climate by measuring the width of the annual rings in tree trunks. Trees grow more when the climate is warm.

Global warming
If the world warms up enough, the ice at the North and South poles could melt. This will unlock vast amounts of water, which may cause the oceans to rise. Coastal cities, like New York City, could be flooded or washed away.

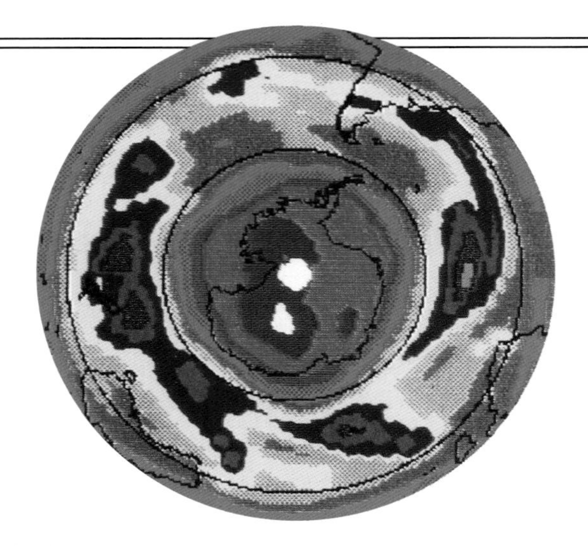

Ozone hole
In the 1980s, space scientists discovered a thinning of the ozone layer *(above)* in the atmosphere over Antarctica. This thinning, called an ozone hole, now happens regularly. If the ozone layer thins too much, it will let through sun rays that harm living things.

FACT BOX

- About 65 million years ago, a huge meteorite crashed down on the earth and caused the climate to change. Many scientists believe that this killed off the dinosaurs and many other species, or kinds, of animals.
- The last ice age came to an end about 10,000 years ago. Geologists – scientists who study the earth – believe another ice age could happen in a few thousand years.
- Mexico City, Mexico, is sometimes so polluted that birds have been known to fall dead out of the skies. Most days, the pollution blocks out the sun.

Acid rain
Acid forms in the air when gases from cars and factory chimneys escape and combine with the tiny droplets of water in clouds. Then, when it rains, the rain is acidic. Acid rain is harmful to the environment and all living things.

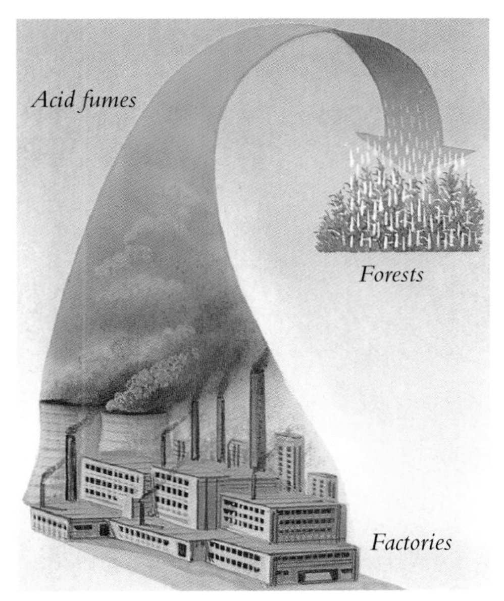

Dirty air
Fumes from traffic and factory chimneys affect climate and air quality. Living things suffer when air quality is poor. Sometimes cyclists have to wear masks when the air is heavily polluted.

RECORDING THE WEATHER

A rocket blasts off from Cape Canaveral, in Florida, carrying a weather satellite called NOAA. *This satellite will circle the earth over the North and South poles and will send cloud pictures and other weather data back to earth.*

METEOROLOGISTS do two main jobs. One is collecting information about what the weather is like now. The other is using this information to forecast what the weather will be like in the future. Meteorologists collect information about the weather at weather stations scattered around the world. They use a variety of measuring instruments. For example, thermometers measure the temperature, barometers measure the air pressure, and hygrometers measure the humidity of the air. Weather vanes show wind direction, and anemometers measure wind speed. Meteorologists also use space technology to help them. They make use of weather satellites that have been sent into orbit to take pictures of clouds and measure temperatures and other weather conditions. Satellites are useful because they record weather in remote regions where there are no weather stations.

Weather balloon
Meteorologists are about to launch a weather balloon called a radiosonde. It will carry instruments high into the atmosphere.

Measuring sunshine

This instrument *(left)* is a sunshine recorder. The glass ball acts as a lens and, when the sun shines, it burns a piece of paper underneath. In this way, the instrument records the length of time the sun is out.

Measuring wind

This small weather station has an anemometer (to measure wind speed) and a weather vane (to show the direction of the wind) mounted on poles above the roof. Note the wispy cirrus clouds in the sky.

Weather at sea

This is an automatic weather buoy. It carries a variety of instruments. Readings are transmitted automatically to weather stations or satellites passing overhead.

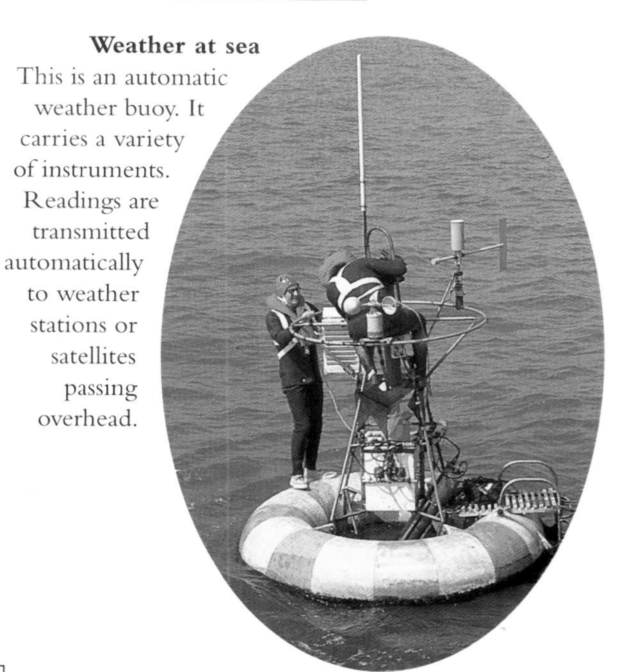

Measuring temperature

Here, a meteorologist is reading thermometers that are in a special enclosure called a Stevenson screen.

YOUR WEATHER STATION

YOU can set up your own weather station at home with a few simple instruments. You will be able to use some of the things you made earlier in this book, such as the weather vane, hygrometer, and rain gauge. In addition, you will need a thermometer and a barometer, which both can be purchased inexpensively. Pick up some pine cones in the park and some seaweed at the beach.

Make a note
Take measurements with your weather instruments every day and write them down in a notebook. Also, make a note of what the weather is like generally – clear, cloudy, rainy, frosty, and so on. Remember to write down the dates!

How humid is it?
Your hygrometer will help answer this question. Note the position of the pointer on the scale. When it tilts up, the air is moist, and rain could be on the way.

Which way is the wind blowing?
Remember, the arrow points in the direction from which the wind is blowing. A north wind blows from the north.

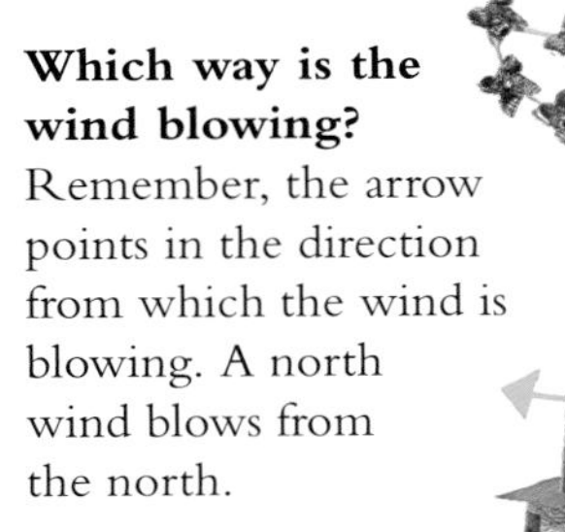

How hot is it?
These thermometers at a weather station measure the temperature in the shade. Make sure you place your thermometer in the shade.

How much did it rain?
A rain gauge indicates how much it has rained. Measure the amount of water in the jar with a measuring bottle. Be sure to empty the jar after you have finished.

Red clouds in the morning provide a warning of stormy weather to come.

The puffy white cumulus clouds tell us the weather is going to stay calm.

PLANT POINTERS

No home weather station would be complete without pine cones. When they are ripe, pine cones open on dry days to shed their seeds. They close up if the weather is humid, or damp. Even after they have shed their seeds and fallen, pine cones still tend to open wider on dry days and close on damp days. Seaweed also changes as the humidity changes. If the weather is dry, the seaweed feels dry and brittle. If the weather is humid, the seaweed feels flexible and damp.

Read the clouds

The kinds of clouds in the sky often give you a clue about how the weather is going to develop during the day.

FORECASTING

To prepare weather forecasts, meteorologists traditionally draw a number of weather charts, or maps. At certain times of the day, they gather all the latest information about the weather in their regions and plot it on a chart in the form of symbols. They compare this chart with one they drew some hours before, which gives them an idea about how the weather is changing. They draw another chart that shows what they think the weather will be like in the near future. Using this chart, they issue a weather forecast. Meteorologists increasingly use computers to help them forecast. More easily and accurately than people can, computers trace patterns in past weather and show how the weather should develop. However, you cannot always rely on these methods – even computers get it wrong sometimes.

Satellite weather
A weather satellite called *GOES* sent back this picture *(above)* of North and South America. It shows swirls of clouds scattered over the two continents and the oceans that surround them.

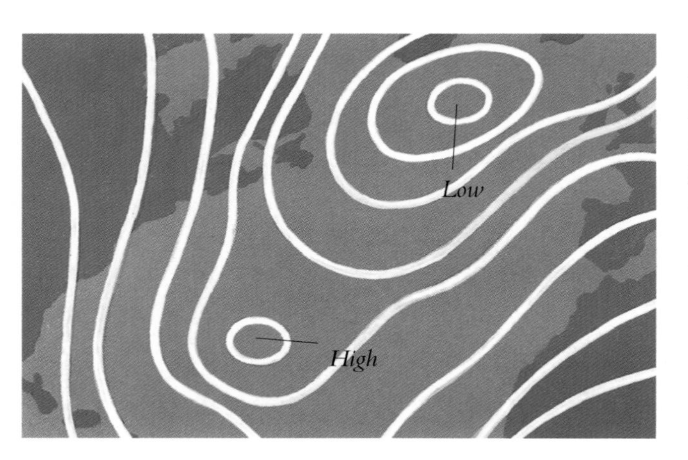

The white lines on this weather map (left) *of the Atlantic Ocean are called isobars. They connect areas that have the same air pressure. The center of an area of low pressure is marked low. The center of a high pressure area is marked high.*

This map (right) *shows the British Isles and northern France. Again, the white lines are isobars.*

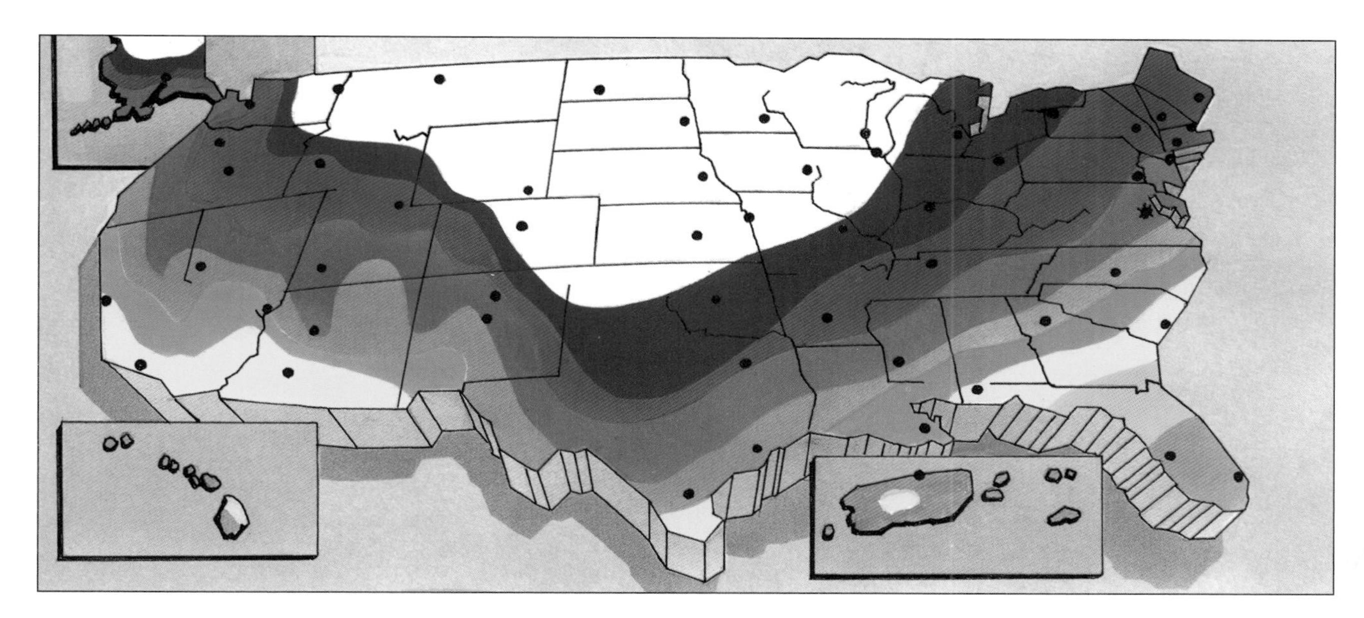

U.S.A weather
Newspapers often carry simplified weather maps such as this one *(above)*, which shows the weather expected over the United States for a day in January. The different colors indicate different temperatures: white is coldest, and orange is warmest.

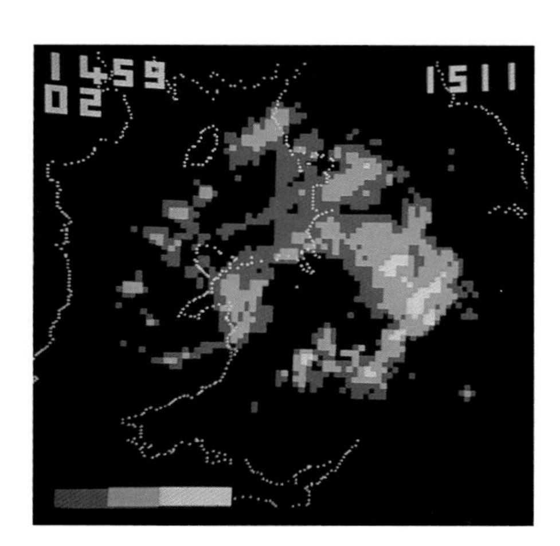

Radar scan
Meteorologists use radar to detect rain. This weather radar scan shows showers taking place over North Wales in the United Kingdom. The yellow areas show the regions of heaviest rainfall.

FACT BOX

- Weather satellites can take pictures of the cloud cover over the earth at night as well as during the day.
- The World Meteorological Organization (WMO) is the main international weather organization. Weather forecasters send information about their local weather to the WMO. In return, they can gain information about the weather in other parts of the world from the WMO.
- The chart weather forecasters prepare that indicates the weather that occurred at a certain time is called a synoptic chart. The chart they prepare that indicates the weather forecast is a prognostic chart.

GLOSSARY

anemometer – the instrument used in weather forecasting to measure the speed and force of the wind.

atmosphere – the layer of air and gases that surrounds the earth.

atmospheric pressure – a measure of the weight of the air pressing down on the surface of the earth.

barometer – the instrument used in weather forecasting to measure atmospheric pressure.

Beaufort scale – a range of code numbers that describes the force of the wind, from 0 (calm) to 12 (hurricane), that goes along with various wind speeds. The Beaufort scale is named after British naval officer Sir Francis Beaufort, who introduced it.

Celsius – a scale for measuring temperature where freezing is at zero (0) degrees and boiling is at 100 degrees. The scale was introduced by Swedish astronomer Anders Celsius.

drought – a long time without rain when living things do not have the water they need.

equator – the imaginary circle around the middle of the earth between the North and South poles that identifies a region with a climate that stays hot all year round because it always gets the greatest amount of direct sunlight as the earth turns on its axis.

Fahrenheit – a scale for measuring temperature where freezing is at 32 degrees and boiling is at 212 degrees. The scale was introduced by Gabriel Daniel Fahrenheit.

front – the point, or boundary, where two air masses that have different temperatures and different amounts of moisture meet. When two fronts meet, there is a change in the weather.

humidity – a measure of the amount of water, or moisture, in the air.

hygrometer – an instrument used in weather forecasting to measure humidity.

isobar – a line on a weather map that connects areas with the same atmospheric pressure.

land breeze – a gentle wind that blows from the land toward the sea.

meteorologist – a person who studies the science of meteorology (weather conditions) and forecasts and reports on the weather.

meteorology – a science that concentrates on the atmosphere, climates, and weather conditions in regions around the world.

monsoon – a wind that reverses its direction in winter and in summer. Monsoon winds commonly affect southern Asia around the Indian Ocean, often bringing heavy rains in the summer season.

ozone – a kind of oxygen gas that exists in a layer of the earth's atmosphere that blocks dangerous sun rays.

precipitation – any form of water (rain, sleet, hail, or snow) that comes out of the air and falls to the ground.

prevailing winds – the winds in a large area or region of the world where the air movement almost always travels in the same direction.

prognostic chart – a weather map prepared by forecasters to show information about the weather that a certain area probably will experience at a particular time in the future.

prominence – a fountain of fiery gas that shoots out thousands of miles (kilometers) from the surface of the sun.

radiosonde – a scientific instrument carried high into the air, usually by a weather balloon. The radiosonde sends back information to earth about the atmosphere and weather conditions.

satellite – Scientific equipment launched from earth to outer space that sends back information to help scientists understand how the sun affects the world's climates and weather.

savanna – a huge grassy region with hardly any trees in a tropical climate. The region experiences heavy rain for part of the year.

sea breeze – a gentle wind that blows from the sea toward the land.

spectrum – the spread of colors found in light in the order red, orange, yellow, green, blue, indigo, and violet. A spectrum appears when white light passes through a prism (a transparent solid object such as a wedge of glass) or through raindrops to form a rainbow.

static electricity – bits of electricity that stay in place and create an electric charge.

stratosphere – an upper layer of the earth's atmosphere, above the clouds.

sunspots – dark areas that appear on the surface of the sun and are a little bit cooler than the temperature on most of the sun's surface.

synoptic chart – a weather map prepared by forecasters to show information about the weather that a certain area is having at the present time.

transpiration – the process in the water cycle by which plants release water vapor into the air. The vapor will eventually turn back into liquid and fall back to the earth as some form of precipitation.

troposphere – the layer of earth's atmosphere closest to the surface of the earth, where clouds form.

tundra – a huge treeless area in the Arctic regions that has very long, harsh winters and where the ground beneath the surface is always frozen, even in summer.

BOOKS

Blizzards. Steven Otfinoski (TFC Books)

Discover Weather. Robert W. Grumbine (Forest House)

Hurricane! The Rage of Hurricane Andrew. Patricia Lantier-Sampon and *Miami Herald News*

Hurricanes: Earth's Biggest Storms. Patricia Lauber (Scholastic)

It's Raining Cats and Dogs: All Kinds of Weather and Why We Have It. Franklyn M. Branley (Avon)

The Nature and Science of Rain. Exploring the Science of Nature (series). Jane Burton and Kim Taylor (Gareth Stevens)

Rain: Causes and Effects. Phillip Steele (Watts)

Simple Weather Experiments with Everyday Materials. Muriel Mandell (Sterling)

Storms: Nature's Fury. Jenny Wood (Gareth Stevens)

Thirty-Nine Easy Meteorology Experiments. Robert W. Wood (TAB Books)

Thunderbolt: Learning about Lightning. Jonathan D. Kahl (Lerner Group)

Tornadoes. Charles Rotter (Creative Education)

Weather Report (series). Ann Merk and Jim Merk (Rourke Corporation)

Weather Spotters Guide. Francis Wilson and Felicity Mansfield (EDC)

VIDEOS

Exploring Weather (series). The Atmosphere in Motion. The Job of a Meteorologist. (United Learning, Inc.)

Flash to Bang: Lightning. (United Learning, Inc.)

Meteorology. (PBS Video)

Weather: How Do Clouds Float? (Agency for Instructional Technology)

The Weather People. (Barr Media Group)

What Makes the Weather? (Film Ideas, Inc.)

WEB SITES

www.nwlink.com/~wxdude

www.whnt19.com/kidwx/index.html

Some web sites stay current longer than others. For further web sites, use your search engines to locate the following topics: *clouds, humidity, meteorology, rainbows, storms, sun, weather, and wind.*

INDEX

PICTURE CREDITS

b=bottom, t=top, c=center, l=left, r=right

Zefa: pages 4br; 20tl; 24tl; 25bl; 37tr; 40b; 43br; 51tr; 55tr. Trip: pages 5bl; 8bl; 33tr, c; 37bl; 42tr; 49bl; 53br, c; 57br; 60bl. Tony Stone 7tr; 11bl; 13bl; 17tr; 27br; 31tl; 40tl; 41b; 48c; 51br; Bruce Coleman: pages 36c; 50c; 51tl; 52br; 56bl. Spacecharts: pages 10bl, c; 11t; 24c; 57tl; 58tl; 63tl. Robin Kerrod: pages 4tr; 12tl, b; 13tr, 16; 17br; 21tr; 28tl; 32; 33bl; 34tr; 41tl; 49; 50tl; 52tr; 56tr; 59tr; 61. National Meteorological Library: 58br; 59tl. Mary Evans Picture Library: 36tl. A.J. & J. R. Chambers 36br. Mansell Collection 5br. Papilio Photographic: 52bl.